The Flying Mystique

HARRY BAUER

illustrated by Kaaren Shandroff

The Flying Mystique

exploring reality & self in the sky

DELACORTE PRESS/ELEANOR FRIEDE

Published by
Delacorte Press/Eleanor Friede
1 Dag Hammarskjold Plaza
New York, N.Y. 10017

ACKNOWLEDGMENTS

Excerpts and chart on pages 83–84 adapted from *The Psychology of Consciousness*, Second Edition, copyright © 1977 by Robert E. Ornstein. Reprinted by permission of Harcourt, Brace, Jovanovich, Inc.

Lines from "Flying Above California" from POEMS 1950–1966: *A Selection* by Thom Gunn. Used by permission of Faber & Faber Ltd.

From *Moly and My Sad Captains* by Thom Gunn. Copyright © 1961, 1971, 1973 by Thom Gunn. Reprinted by permission of Farrar, Straus & Giroux, Inc.

Manufactured in the United States of America

First printing

Designed by Terry Antonicelli

Library of Congress Cataloging in Publication Data

Bauer, Harry.
The flying mystique.

Bibliography: p.
1. Flight. 2. Private flying. I. Title.
TL710.B37 629.132'5217 79–27690
ISBN 0–440–02722–5

TO JANE

Sometimes

on fogless days by the Pacific
there is a cold hard light without break

that reveals merely what is—no more
and no less. That limiting candour,

that accuracy of the beaches,
is part of the ultimate richness.

THOM GUNN,
"*Flying Above California*"

Contents

Lilienthal's glider 1896
Orville WRIGHT 1909
Spirit
St. Louis
Harry Graham
Captain

of
Louis
Wilbur WRIGHT
Capt. CHARLES A. LIN
BERGH-192
Donald Hall
Edwards

Prologue

ALL human inventions present us with unavoidable questions. If the new invention is to be used, whether it is a Paleolithic axe, a nuclear reactor, or the airplane, how will it change other aspects of existence? Will the changes be trivial or profound? The airplane has become part of our contemporary world, but all the ways flying might touch those who actively make the airplane an extension of themselves are not yet certain. What might it hold for those who are willing to detach themselves from familiar routines, trust their lives to their own abilities, and then face the implications of their independence when they return to earth? The opportunity to choose the answers to these questions is still open to us.

Although the sight of an airplane overhead is an ordinary part of our daily life in America, flying itself is a new experience for man. Forms of life can be traced back millions of years, but it was only in 1903, about a single life-span ago, that the Wright brothers first flew at Kitty Hawk. Flying is a new magic—nothing in man's history has been quite like our freshly sprung ability to actually leave the earth and sail about the clouds whenever we wish. The technology of flying has thrust us quickly from the Wright Flier to the Concorde and beyond, but when set against the backdrop of all of man's past, we haven't had very much time to consider what this ordinary magic of flying might be and what it might mean for us.

Flying holds an important place in the mythology and value system of American culture. For one thing it represents a highly developed and efficient technological society. After all, we can fly safely, comfortably, and cheaply from New York to San Francisco in less time than it took our grandmothers to cook a Thanksgiving turkey. We can enjoy a glass of wine and watch America slip by 40,000 feet below us and marvel at the ease and convenience that technology has brought to our lives. On a human scale the early aviation pioneers are in many ways personifications of what we have traditionally insisted upon in our cultural heroes. The Lindberghs, the Earharts, the Wright brothers, were all strongly individualistic, if not downright loners. Like another folk hero who would also be a misfit in today's corporate society, the frontiersman, they were eager to face new challenges and willing to work hard to achieve the goals they had set for themselves. In setting these goals they created their own worlds, not completely accepting the one in which they found themselves. This restructuring of reality is a vital part of flying and a major theme of this book.

The millions who visit the National Air and Space Museum in Washington, D.C., are drawn there partly by the dual nature of aviation's appeal. The exhibits include Lindbergh's "Spirit of St. Louis," a symbol of one side of the duality, and a collection of rockets, moon vehicles, and space capsules, symbols of the other.

But we don't have to go to the Air and Space Museum to find evidence of aviation's attraction. If we look closely around any general aviation airport, we will surely see a few people just standing around watching whatever might be going on. They won't have any particular business there—they will just be watching. Some may be in cars, finishing a brown bag lunch away from the office. Others may have transistor radios and occasionally sweep the horizon with their eyes looking for an airplane they heard call in on the tower frequency. Some may be bold enough to get out of their cars, walk past the "Pilots and Passengers Only" sign and onto the concrete or grass, and actually touch and look inside the small airplanes that are parked there. Most of these people don't fly themselves, but somehow they enjoy the ambience of the setting they have entered. They are surrounded by complex, well-designed and -engineered machinery that others have worked hard to understand and control. They may also be attracted by something that is difficult to define in our technological culture, something that puts existence together in a different way. Somehow, these elements seem to fit together at the airport, and our watchers find themselves there.

I live in northern California, where the weather, scenery, and cities make an ideal location for the filming of movies and television series. People tend to gather in places where filming is going on, and I believe they gather for some of the same reasons that others are drawn to airports. In general the crowds around a film location are well-behaved and cooperative. They want to get a sense of the action, but they don't want to interfere. Some of them, of course, hope to get a glimpse of someone famous, but that's fantasy and part of the airport syndrome also. The crowds are really there to see how it is all done. How does the combination of cameras, lights, props, and actors make a movie? How does someone put all these pieces together? How does a sense of how it might come together combine with a technical understanding of each minute part? How does knowledge of focal length, color balance, and acting technique add up to a film that will make us laugh, cry, or buy Coca-Cola? The crowds know, as you and I know, that a good film will combine technical expertise with an approach that will put the pieces together into something that is greater than the sum of its parts. We have all seen movies that approach technical perfection but that have nothing to say, that bore and tire us. We have also seen movies that have some grand vision of existence that we can't comprehend because the technical quality is so crude that even our sense of imagination and fantasy won't override the awkwardness. Neither kind is any good.

It's only when a movie combines technical command with a vision of how it all goes together that we leave the theater or television screen with a feeling of satisfaction. Let's call this notion of how things can go together "holistic" and use the term to refer to a nonspecific, all inclusive approach to the universe. The holistic approach does not assume that there is a single correct answer to any particular question, but that there may be an infinity of answers. The validity of the answer depends on how we use all the capacities and awareness we have to discover it.

Now we are getting closer to what flying is all about—or at least what I think it is all about. Flying combines the technical and the holistic aspects of life. It is an activity that forces us to put together and to relate many parts of our lives that didn't seem related before along with a technical understanding of what is physically happening to the airplane. That's why flying can be so beautiful. We get a grand view of the universe at the same time that we get a closer look at its parts.

Flying demands a great amount of skill and a command of a complex body of technical information. The pilot with superb flying skills but with no knowledge of aerodynamics or of the limitations of his airplane is a threat to himself and everyone else. So is the pilot who has mastered the theoretical aspects of aviation but who can't control his airplane. Each aspect depends on the other, and neither is sufficient by itself. It's the combination, the way that the parts are brought together, that is critical.

Flying is similar to many other complex activities in its need for this combination of holistic and technical or intellectual skills. Athletics, art, music—all require a similar combination of holistic and specific aspects. The musician needs to understand the complexities of harmony and technique while he interprets someone else's music or composes his own. Also, these activities are ways that people find to explore the unknown capacities that lie within themselves. Each demands that talents which may have been hidden or denied be discovered and developed. The music student who finds out that he has perfect pitch and the artist who learns how to use color in a unique way are two examples. The skier who learns to accept his own fear and to recognize its value is another.

In all cases we must invest our total identity into the activity if we are to find the magic it may hold for us. If we are to reach the delight that comes with the joyful exercise of well-developed skills, we must practice, we must work, we must occasionally fail, and we must endure times of complete frustration. Unless we wish to be mere dilettantes, this is the price we have to pay. Those who do find themselves pay willingly.

But there is a subtle danger in this that you must be aware of from the beginning. The experience of art, music, or flying is going to change the way you perceive the universe. The world of the dancer is not the world of his audience. If you decide to fly, the reality you return to will not be the reality you leave below when you take off for the first time. You must be willing to accept the anxiety and excitement that this will cause you.

These are the ideas of this book. The most appropriate metaphor we can use to discuss them is flight itself. All flying can be reduced to four basic maneuvers: straight and level flight, the climb, the turn, and the descent. Every flight maneuver either is one of these alone or is some combination of them. The thoughts we will deal with seem to cluster themselves around the maneuvers quite readily, so we can examine each maneuver as we explore the notions related to it. Just remember that a metaphor is only a convenient device for relating the unfamiliar to the familiar. It is meant to be a guide, and its value evaporates when overextended.

I. The Climb

Preparing to Fly

EVERY flight begins with a climb, an escape from the earth to some safe altitude where the airplane can find its natural environment. When we are ready to start the climb, we will see that as the airplane begins to leave the ground during the takeoff there is very little for the pilot to do. The airplane, formed for movement through the air, lifts effortlessly off the runway.

Because part of flying is very rational and orderly, we have many things to do before we are ready to fly. The sensible pilot approaches flying with the rigor of a scientist conducting an experiment. For an experiment to be valid, all the factors that influence it must be strictly controlled and the conditions of the experiment clearly stated and carefully followed. Order is the essence of science, and if any element other than the one under observation is left uncontrolled or is beyond the scientist's ability to control, the result will not be science and order, but chaos.

The goal of the pilot's experiment is a safe flight. He knows that if all the factors that influence his flight are controlled, if all the procedures he has learned are followed, the result will be a safe landing at his destination. He also knows that if anything critical slips from his control the result might be disaster. So the pilot acts in the finest traditions of Western thought, applying logical, rational, and sequential procedures to eliminate danger and chance. He follows a set procedure when inspecting his airplane on the ground before flight to ensure that every vital part and system are in sound condition and are operating properly. He already has calculated how long it will take to get to his destination and has learned what the weather will be along his route of flight. He knows how much fuel the engine will burn in a given period of time and actually looks in the fuel tanks to be sure there is enough. If the weather in any way approaches the limits of his ability to fly safely, one of the elements affecting the outcome of the experiment would be beyond his control, and he will cancel his flight and remain on the ground. When it is time to start the engine, he uses a checklist to make sure that all the steps required are done in the proper order. When the engine is running, he checks the instruments for indications that it is performing as it should. If he doesn't have a written checklist at hand, he follows procedures that have been committed to memory, aided by easily remembered acronyms and phrases. The Federal Aviation Administration recommends "CIGARS":

C—Controls free
I—Instruments checked
G—Gas switched to fullest tank
A—Altimeter set
R—Runup complete
S—Safety

A pilot with a greater sense of whimsey might prefer "Can I go flying today, Peter Rabbit?"

Can—Controls free
I—Instruments checked
Go—Gas switched to fullest tank
Flying—Flaps set
Today—Trim set
Peter—Prop set to highest RPM
Rabbit—Runup complete

In each case a carefully established procedure has been followed, and the pilot is satisfied that his airplane is safe and ready to fly.

When we have done all these things, we are able to begin our flight, to leave ordinary reality. Our perceptions are challenged by several things before we take off. The noise you heard coming from the loudspeaker above your head is the controller in the tower giving instructions and information you need to know about the wind, traffic, and air pressure. Don't be alarmed if you aren't able to decipher the words, let alone determine their meaning. I have a friend with an extraordinary gift for words who can't suppress giggles every time he hears the strange cryptic language that pilots and controllers use when they talk to each other. Your ear will soon be able to decode the noise and understand the words, and you will soon learn their precise and unambiguous meanings. It is as simple as adding new words to your vocabulary. Later you will be able to put those words into new patterns of meaning so that if a controller says, for example, "Enter on right base for runway 28. Traffic on one mile final," you will immediately have a three-dimensional image of the situation, and your eyes will automatically turn in the direction of the other airplane. You will perceive the relationships among time, space, and distance in a way that you aren't able to do now.

Now let's put a new meaning on a familiar symbol. Sitting side by side in this small airplane is something like being in a sports car. We even have something that looks rather like a steering wheel in front of each of us. At first some people make the familiar association and try to steer the airplane on the ground with the wheel, as they would an automobile. It won't work. On the ground we steer the airplane by the rudder pedals with our feet, the way we did with carts when we were children. This may seem clumsy and primitive until we remember that airplanes are awkward and ungainly machines on the ground: their shapes were formed for movement in the air.

Once we have made some final checks of the engine and the controls and have been cleared by the tower, we can point the airplane down the runway and take off. All we really have to do to keep it pointed in the right direction is to apply pressure on the rudder pedals. As the speed increases the control surfaces on the wing and tail assembly become more effective, the shape of the airplane begins to do its work, and we lift gently into the air.

Two Ways of Perceiving

NOW we have established the climb, and are flying in a straight line as we increase our altitude. The airplane has been trimmed—adjusted—so that it will continue this way even if we take our hands and feet off the controls. All the forces of thrust, drag, lift, and gravity are balanced, and we are in a state of equilibrium, neither accelerating nor decelerating, just climbing at a constant speed. The airplane is simply following the laws of classical Newtonian physics. While we are in the climb we can speculate on what meaning flight may have for us in a personal sense. Of course, if this were a real flight, we would be so completely involved in the flight itself that such a dialogue would be impossible. Total absorption in the here and now is one of the joys of flying, and we will examine that more carefully later. For now, and since this is a vicarious journey that we've set upon, we can afford the luxury of introspection.

We discussed how the careful pilot acts as a scientist conducting an experiment, carefully examining each piece of vital data that will influence its outcome. While we are in the air we must experience the universe in another way. Now we have to deal in great patterns of information, patterns that are constantly changing and that can't be perceived in a logical, sequential manner. For example, as we are climbing to cruising altitude we are moving through three-dimensional space, and what is important is our relationship to other objects in that same space. How far away are those hills that lie ahead of us? If we continue at this rate of climb, will we clear them at a safe altitude? We are forced to perceive the universe in a holistic way, an experience for which most of us are not prepared.

There seem to be some reasons for this difficulty, reasons not directly related to flying. They have to do with heredity and environment. "Environment" has a rather limited meaning today, usually referring to trees, birds, freeways, and air pollution. We are using the term here in the broadest sense, to include everything we perceive and experience.

Part of the equipment that we are all supplied with is the mysterious bundle of tissue inside our skulls, the human brain. Some of the most fascinating scientific research going on in this decade includes investigations into how the right and left hemispheres of the brain use different modes of thinking and of perceiving information. This split brain theory is a handy tool for interpreting human behavior and cultural differences. It can be used to explain why some people prefer Mozart to Fleetwood Mac, the World Series to the Superbowl, and why the Occidental and the Oriental often have trouble comprehending each other. It also offers some clues as to why flying can be such a satisfying and fulfilling experience.

Like our scientist/pilot, the left hemisphere seems to operate in a logical, rational, and sequential style. Our verbal and analytic capacities appear to be located there, as well as our ability to use numbers. It's the left side of the brain that enjoys a baseball game, with its step-by-step progression through nine innings and its flood of statistics that allows us to relate Joe DiMaggio's batting average to Vida Blue's earned run average.

The right hemisphere is less rational and more intuitive. It deals best in patterns of information, in relationships rather than specific bits of data. Using a holistic approach, it helps us to determine relationships between the parts of a whole and to perceive spatial relationships. Football players and their fans are good at this. They can tell when and where a downfield receiver will be in the clear and how high a pass must arch to be there when he is. Studies also suggest that the right hemisphere is the source of musical and artistic abilities.

In Western civilization we put the highest value on left hemisphere thinking. We stress logic, reason, and analysis in nearly every aspect of our lives. Our schools, with their sequence of grades and divisions of subject matter, certainly reflect this. Indeed, science itself, the application of reason in the search for knowledge, is a Western invention. Robert Ornstein, in *The Psychology of Consciousness*, points out this cultural bias and uses examples from the Sufis and other non-Western groups to show that traditional Eastern civilization tends to place greater stress on holistic, right hemisphere, modes of thinking than we do in the West. None of the writers currently dealing with these ideas suggests that we in the West abandon science and seek gurus on mountaintops, but they generally conclude that we ignore our right hemisphere potentials at our individual and social peril. There seems to be something terribly wrong in not recognizing and developing such a precious resource of human potential.

When we analyze flying, we see that it uses a combination of both modes of thinking, the rational and the holistic. The pilot must use both to survive in the air, and he must integrate the results. Neither side of his brain is ignored or allowed to remain idle. Both are used to the fullest, and no potential is left untapped. Let's put our left hemispheres to work to analyze this notion and to examine it more closely.

Since we have already discussed briefly the rational, sequential aspect of flying, let's look at its holistic aspects in terms of what we are doing at this moment, keeping the airplane in a climb. We are trying to make sense out of all the messages that are impinging on us at the same time, and it's this ability to integrate and perceive patterns in vast amounts of diverse information that is associated with the right hemisphere of the brain. In the climb, to be sure that we are maintaining a straight climb with the wings level, we are watching the airspace in front of us to be sure it is empty, the angle of the nose to the horizon (pitch), and the position of the wingtips. We are also scanning the instruments to see if we are holding the proper airspeed as we climb and that our vertical speed is adequate, and we monitor the gyrocompass to be sure that we are holding the heading we want. Our hands and feet react to subtle differences in control pressures and apply whatever corrections are necessary to hold our desired pitch, altitude, and direction. When we encounter sudden updrafts and downdrafts, our entire body signals us that the force of gravity is apparently changing, and we make the appropriate control changes. This is the feeling we experience in a fast-moving elevator. The sound of the engine will change during the climb if the speed of the airplane changes and the pitch attitude of the nose varies, and our ears provide another set of messages for us to process and consider. Somehow out of this kaleidoscopic rush of information we are able to discern a sensible pattern and keep the airplane in a steady climb. It's this need to find patterns in a barrage of changing messages from several senses at once that makes flying a holistic adventure.

Actually we don't switch from one hemisphere to the other when we are flying; we are using them both, blending the special abilities of each for a smooth and safe flight. We demand all the power we can get from the rational, logical side of our brain and all the power we can get from the holistic side as well. Flying gives us a chance to use our skills and intelligence to the utmost, coordinating our minds and bodies to levels of performance that we seldom need to attain in other areas of our lives.

Undiscerned Possibilities

FEW people are lucky enough to find something that demands so much of them and at the same time brings them so much pleasure. Flying isn't for everyone, but those of us who fly know that it gives us the opportunity to exercise our human capacities to the fullest. Saint-Exupéry was speaking of this in *Wind, Sand, and Stars* when he described riding a bus to the aerodrome to make his first flight as an airmail pilot. His fellow passengers were clerks and petty officials on the way to live through another dreary day of sorting and filing in obscure corners of the French bureaucracy. He was stunned by the banality of their conversation and felt a deep sense of sorrow for them and for the stunted lives they had chosen to lead. They were starting a day of lifeless routine and security, straining what little of their potential that hadn't yet withered away, while he was going off to fly. He saw them as sleepwalkers who could never dream of the wonders around them and within themselves. "Now the clay of which you were shaped has dried and hardened, and naught in you will ever awaken the sleeping musician, the poet, the astronomer that possibly inhabited you in the beginning."*

Poets like Saint-Exupéry have means of considering the human condition; so do psychologists who use tests, surveys, and other tools to try to find what might be concealed within us. Recently two American psychologists, Novello and Youssef, reported that they found that pilots, both men and women, have personality profiles that differ significantly from those of the general population. Specifically they found that pilots, as a group, have stronger than average drives to accomplish difficult tasks, to do new things, to talk about personal adventures, to

---

* Antoine de Saint-Exupéry, *Wind, Sand, and Stars* (New York: Harcourt, Brace & World, 1940), p. 23.

argue for their own points of view, and to be interested in the opposite sex. Thus science validates the myth that Hollywood has been presenting to us for years, from *Dawn Patrol* to *The Great Waldo Pepper,* the myth of the pilot as romantic hero or heroine. The study also showed that pilots are less likely to be concerned with orderliness and with doing what others expect of them. Taken together these factors present a strange paradox. The elements that come together to create the pilot can also be the ones that can lead him to disaster.

That the pilot seems to have within him the seeds of his own destruction brings us to another idea that you must consider seriously, but not in a somber or negative fashion. You are a brave person. Not brave because you are going to be facing any physical dangers: you are not really going to. I mean brave in another, deeper sense. By being on this flight you have shown that you are willing to explore your own identity to discover what might lie within you. Your human clay has not hardened, and you are also willing to explore your own perceptions of the universe, knowing that you may be forced to set aside many comfortable and cherished assumptions. The idea that you must approach honestly and directly is that flying very dramatically makes the pilot solely responsible for his own life.

2. Straight & Level Flight

On Being Unobtrusive

Now our climb to altitude is accomplished. To make the transition to straight and level flight, we simply lower the nose of the airplane to the horizon, reduce the power from a climb to a cruise setting, and adjust the controls with the trim tab to hold the altitude. The airplane should now be level at a constant altitude, and we will be constantly sweeping our eyes across the sky and the instruments to be sure that it is staying that way. Only very light and subtle pressures on the controls will be necessary if we stray slightly from our heading and altitude.

Just as most driving is in a straight line, most flying is straight and level. We simply steer the airplane on a constant heading and maintain a constant altitude. Although this is the least complicated of all flight maneuvers, many pilots never quite get the knack of it and as a result find altitude and heading control a difficult chore. Probably their trouble comes from trying to do too much themselves and not letting the airplane fly itself.

In straight and level flight the forces acting on the airplane are again in balance. The thrust provided by the propeller, or jet engine, is equal to the drag, the resistance of the air to the airplane's motion through it. At the same time the lift created by the air passing over the wing matches the total weight of the airplane, and we can maintain a constant altitude. Lift is created by the wing's airfoil, its shape when we look at it from the side. The airfoil simply causes a difference in the air pressure between the top and bottom surfaces of the wing as the air flows around it. The pressure on the top surface is reduced, and the pressure on the bottom remains relatively constant. The greater pressure on the bottom surface provides the force called lift that is equal to the weight of the airplane and sustains it in the air. A simple way to illustrate how this works is to blow across the top of a sheet of paper held to the lips. The paper will rise as the air passes over its surface.

When these forces are in concert, the pilot's task is merely to monitor the situation and to intrude as seldom as possible. Flying becomes an academic exercise, another illustration of Newton's laws of motion and equilibrium, a collection of impersonal physical forces that seems external to us and part of an apparently mechanical universe.

But it is an abstract, academic exercise only if we are able to ignore what our situation really is as we pass through the air several thousand feet above the surface of the earth. We are existing for the moment in an alien environment. Unlike birds, insects, and a few other living things, man has not evolved to survive in the air. The same air that we depend on for life itself will not sustain the weight of our bodies. We can't soar and float as an eagle; we fall through the air, our speed accelerated by gravity until we crash to the ground.

Although his body alone won't allow him to fly, man has a greater gift: his ability to create tools that manipulate parts of his immediate environment and that expand the ways in which he can survive in the larger environment in which he lives. The airplane is simply another invention that allows man to project himself into the world. Just as the wheel and the boat let him extend himself across the surface of the earth, the airplane lets him extend himself into the sky above it.

There is a tendency in our culture to assume that since man has invented a machine that flies he has conquered the air; that the Master of the Earth has also become the Master of the Sky. This is not true. The notion flows from the false assumption that man is at odds with his environment. The airplane does not "conquer" the sky any more than the igloo "conquers" the Arctic. Both unobtrusively use the properties of their environment, accommodating to it. The assumption also views man as a creature apart and separate from his physical environment, not part of it. Our habit of thinking in linear and sequential terms also leads us to perceive our situation in a framework of cause and effect. Man, "A," directs his airplane, "B," against the atmosphere, "C," and the result is flight, "D." This mode of thinking is certainly useful and important but becomes absurd when applied beyond its limits, as Rube Goldberg so delightfully showed us. In our case the airplane simply cooperates with the properties of the atmosphere to achieve flight. Nothing is wrested from an unwilling universe. We are not standing outside of nature, forcing it to do our bidding once we have learned the laws that govern it. We use the air to fly, but the air is the same after we use it as before, just as the wake of the ship crossing the ocean quickly disappears, leaving the ocean as it was before the ship passed. It is only when man abuses some aspect of his environment and his relationship with it is thrown askew that nature will seem to turn against him.

We become aware of this, as pilots, when we fly near large cities with their yellow-brown blankets of smog and pollution hanging over them. Air pollution is such a common feature of urban life today, and most of us are urbanites, that we tend not to notice it until it becomes particularly foul. Some of us have become so accustomed to brown skies and haze that it all seems perfectly normal. Occasionally, depending on where we live, we are told on certain days that the air is so polluted that we should avoid outdoor activities or even stay inside if we can. This too becomes the norm, and we accept it as an inescapable part of life. Sometimes a storm sweeps through, and for a day or two the air is clean, and we comment on its clarity and its beauty. But soon, little by little, the pollution returns and life is "normal" again. The stinging in our eyes and the subtle tax on our respiratory systems are no longer perceived.

One of the factors that contributes to the concentration of pollution in the atmosphere is a phenomenon called an inversion layer, a layer of air several thousand feet above the ground that is warmer than the air beneath it. Since warm air expands to fill a greater space than needed by the same amount of cool air, the warm air squeezes the cool air in its location near the ground. The air nearest the ground then collects the pollution that we pour into it, and the concentration of pollution grows because the air that contains it has nowhere to go.

The pilot flies through the layers of pollution into the air above and sees the sharp line between the clear air and the smog. The tops of mountains in the distance are clear and distinct, the clouds on the far horizon are crisply defined against the blue sky. It takes only a few minutes or even seconds to reach this part of the sky from the ground, and the impact of the different world we have entered is sudden and dramatic. What happened to our ordinary world of brown air and ghostly shapes in the middle distance?

At any rate here we are in clear air, flying in straight and level flight. When we learn to relax, to control our reactions and to manage our coordination, our body will respond intuitively to the subtle changes in the atmosphere and keep the airplane on course and at the altitude we want to maintain. As we do this all our sensory systems seem to operate at peak levels of alertness. Our eyes, our ears, our kinesthetic senses are all responding to the movements of the airplane. They seem to work automatically to monitor our way through the air, keeping us on our invisible path. It is as if we have become part of a complex feedback loop, ourselves and the airplane, holding it to its course, correcting the smallest deviation in heading and altitude. This is a kind of paradox because only when we relax our senses do they perform so well for us.

What seems to be happening is that we are experiencing the world directly and immediately. We don't stop to think about what we are going to do; we have trained our mind and body to perceive and to act without conceptualizing what the action will be. We have for the moment bypassed our rational thought processes through which we usually filter experience and have responded directly to the changing atmosphere surrounding us.

Our habit of abstracting perceptions rather than simply experiencing them is deeply set and difficult to overcome. The pattern of screening the world through abstraction and conceptualization, not experiencing it directly, is so ingrained that we are apt to apply the filters even when it would be easier and more advantageous for us not to. But new habits are difficult to learn, and we persist in old ones without even realizing that they exist.

Once there was a fire in the curriculum center of the school where I taught. The center was a large room created by knocking out the walls between several classrooms and building a roof over the patio they had opened on to. It was used as a workspace for teachers. Each department had its own separate area where books, materials, and records were stored. It was a confused, crowded maze of poorly arranged cabinets, bookcases, and tables, and in addition to being the place where we did much of our work, it was the center of the faculty's socializing. It had been the scene of many important professional and personal interchanges. After the fire destroyed almost everything in the room it was sealed, and no one was allowed in except administrators and department chairmen. I managed to sneak in to look at the ruins before chains were placed on the doors.

While the room was left in this state for a week or so before cleaning and renovation were started, work schedules and social patterns were shattered, and there was a great deal of grumbling about when, if ever, things were going to be put right again. The principal, partly in an effort to end the grousing, decided that the faculty should see how much damage had been done. Rather than arrange for us to actually enter the room to see for ourselves, he had one of the department chairmen take a series of color slides that were shown to us at a faculty meeting one day after classes. Of course, the pictures gave only a limited impression of what had really happened. Besides being of only middling quality, they were a shallow two-dimensional representation of the totality of the fire and the damage it had done. We were simply shown a few details of charred tables, blackened walls, and lumps of plastic that had once been coffee makers and telephones. There was no way to tell how the fire had spread or where it had been the most intense, which had been obvious to me after only a few seconds of trespassing, or how the place smelled of the mixture of melted plastic and smouldering paper. The world had been put at arm's length, flattened, and made acceptable—put into a format that most of us could be relied on to comprehend. The symbols of reality had been substituted and accepted as sufficient when the reality itself was at hand, and hardly anyone felt that this was an imperfect way to learn about the fire. Most of my colleagues were satisfied that they "knew" what had happened.

After several months of planning and evaluation the curriculum center was remodeled and outfitted with new furniture and equipment. The room was even carpeted to enhance the color coordination and to provide a sound-muffling device, and the place took on the appearance of the lobby of one of the savings and loan offices we see so often in television commercials.

When we fly we must not intrude on our delicate balance of perceptions and reactions, or we will disturb the equilibrium we have established. Once we have trained our responses to the point where they do this for us, we must learn to stand aside and let them do their work without interference. In a sense we have entered a different level of consciousness, very similar to a form of meditation, if indeed it is not the same thing. In the Eastern traditions an individual concentrates his attention on an object, sound, or rhythmic motion in order to reach this state. The concentration opens awareness at the same time that conceptualization is reduced. The mind becomes empty but the person who has reached this state is able to experience directly everything his senses tell him is going on around him. Please don't be alarmed if this sounds too esoteric or ethereal to have anything to do with something that seems as realistic and objective as flying. In fact it is supremely practical. Some of the most advanced meditation techniques were developed by ancient Japanese warriors who needed to be able to respond without hesitation to the movements of their opponents. To hesitate, to stop to analyze, to screen a perception through the thinking process could mean death. In a way our flying becomes a mantra, the focus of our concentration, and we control the airplane without consciously trying to do so.

Others have found that this kind of relaxation of control works for them in different kinds of activities that demand superb coordination of mind and body. Timothy Gallwey in *The Inner Game of Tennis* calls it "Self 2," the mental state of the tennis player that allows his body to become fluid, to let it meet the ball with the racquet to return a shot over the net with grace and accuracy. "Self 1," with its prescribed rules of grip, wrist action, elbow control, and follow-through, must relinquish its control of the body and let "Self 2" take over. Gallwey insists, correctly, that the tennis player must have a thorough knowledge and mastery of the elements that lead to precise responses but that this "Self 1" understanding is only useful to a certain point. Once that point is reached, once the elements are learned, the player must let them act on their own, guided on by the integrative powers of "Self 2."

"Self 1" is analogous to the rational function of our mind. It analyzes, defines, and orders our experience and actions. We have learned to rely on it for most aspects of our lives, and we can't exist without it, but we must understand that it is only part of what we are or can become. We must recognize the limitations of the rational function, and at times we must set it aside.

How the World Goes Away

AND now a strange thing happens. It is so delicate and ephemeral that even to recognize it and allow it to enter our consciousness may destroy it. It is that, while flying, the rest of the world disappears. We suddenly have no past or future. Our lives have been erased from our minds. We have only our flight. We are existing completely in the here and now. By increasing our concentration, through the paradox of not concentrating, the mind is empty of everything but the immediate moment of experience; it is open, clear, and fully alert, and at the same time free of everything but the present.

To put it in mechanical terms—which are not really appropriate but which may serve to explain—it is as if the mainspring of our attention has been wound until it breaks, and the energy released by the snap sustains us in what we are doing. If we take notice that the spring is broken, the moment is destroyed, and the tension of our concentration returns.

I have no idea how or why this happens. I do know that, when we take the world with us when we fly, this phenomenon is not so likely to take place. The work of flying becomes more difficult than when we truly leave the world beneath us. When we can't be sustained by the here and now, the immediate present we are experiencing, our bodies won't respond as we want them to, and our minds won't be open to the world that flying can offer to us. Our flying will be awkward, uncoordinated, and self-conscious, and we will miss the richness of the moments we are in the air.

I learned to fly at a time when my personal life was just beginning to recover from a series of upheavals and profound changes. As a beginner my flying skills were so meager that the slightest test of their limits was a potential crisis. I slowly learned that these crises usually happened when I had been reworking in my mind all the real and imaginary personal traumas of the past year. Instead of living in the air I was still living on the ground while I flew, fruitlessly reacting to events and individuals who for the moment were in a world other than mine. I somehow learned to leave my personal life on the ground as a matter of physical survival. It was only later that I began to discover that I truly was in a different world, a different reality, when I was flying.

Although the setting is completely different, this is the same lesson that Carlos Castaneda had such difficulty learning from don Juan, the Yaqui *brujo*. Castaneda was a graduate student of anthropology who had gone to Mexico to gather material for a dissertation on the use of mesquite and other hallucinogenic drugs among the Yaqui Indians. He eventually widened his interest beyond drugs and became an apprentice to don Juan, an old man who agreed to teach him the ways of a sorcerer. Their relationship lasted nearly a decade and was the subject of a series of popular, if controversial, books. A common theme of the books was the problem Castaneda had in setting aside his usual mode of perception so that he could be receptive to what don Juan was trying to teach him. Castaneda was trained as a scientist, and to his rational mind, accustomed to labeling, classifying, and reducing the universe to discrete parts, don Juan's world made no sense at all. Don Juan insisted that Castaneda had to "empty his mind" before any understanding of another reality would come to him. He, of course, at first resisted, dismissing the idea of thinking of nothing as naive and irrational. It was only when he learned to close his mind to his own world that don Juan's world began to exist for him.

The lesson was that Castaneda had to learn to stop talking to himself mentally or, in don Juan's words, to turn off his "internal dialogue." When we talk to ourselves mentally, carrying on our internal dialogue, we simply repeat and perpetuate our ordinary universe. By thinking about it, analyzing it, and trying to understand it, we validate it and ensure its existence. Since it is always with us in our minds, there is no space left for the existence or even the possibility of anything else. When we finish talking to ourselves in this way, our ordinary world is still there for us, just as we left it, ready for us to exist in it as we always have. "Thus," as don Juan told Castaneda, "we repeat the same choices over and over until the day we die because we keep on repeating the same internal talk until the day we die."*

Joseph C. Pearce, who is concerned with the relationship of culture to perception, calls reaching this state of mind "turning off the chatter" of the roof of the brain. In his *Exploring the Crack in the Cosmic Egg*, he maintains that until we learn to do this we can't begin to know ourselves and the aspects of the

---

* Carlos Castaneda, *A Separate Reality* (New York: Simon and Schuster, 1971), p. 263.

universe that exist outside of everyday reality. The "roof-brain chatter" filters out perceptions that challenge the status quo, and the ordinary world is continually being reinforced.

Flying seems to provide a context in which this phenomenon, whatever it might be called, can occur more easily than in other situations. In a small airplane we know that we are truly flying—we really are moving through the air, above the earth, near the clouds. The impact of the experience is powerful and can't be denied or obscured. It is the harsh purity of the experience that overwhelms us and shakes the foundations of our ordinary world.

How different this is from flying as a passenger in an airliner. There everything is orchestrated to deny the fact that we are actually flying. We enter the airplane through a tube directly from a waiting room, so we can hardly be aware that we are entering a plane at all. Once inside, our perceptions are narrowed and limited to the familiar, the ordinary artifacts of our ordinary world. The cabin attendants may be charming and attractive, but we have met their personae many times before in other parts of the society in which we live. Their job is to make the experience of air travel as bland and routine as possible and to create the illusion that we aren't flying at all, that we are spending a passive interlude between airports, snug and secure with Muzac, martinis, magazines, or perhaps all three. When we are airline passengers, there is little or nothing for us to learn about flying, although there is plenty to learn about how culture can filter and manipulate our perceptions of experience. Air travel has been so drained of vitality and wonder that it is simply an arid exercise, a necessary evil to endure if we want to go from one city to another. The persistent and imaginative traveler can insist on a window seat to get some hints that he really is in the air, but it's not easy to neutralize the stage management designed to convince him that he is not. This is not the place to argue the pros and cons of airline travel, but that kind of flying is not the kind of flying we are speaking about here.

Reconsidering the Familiar

THE flying we are speaking of can truly transport us, not just from place to place, but between different levels of experience. We are living in the immediacy of this moment's experience, and now that we are separated from the past and the future, we need to recognize that we have become separated from our own pasts as well. When we leave the world behind, we leave ourselves behind also. Our separation from the earth is much more than a physical separation, although it is that, of course. Our ties with the earth are now only the voices we can hear on the radio and what we can see through the windows. But even the familiar things we see below us have lost their usual meanings since our psychological bonds to the world have been temporarily suspended.

This phenomenon is not mysterious or mystical—it just is.

We know that the same event, object, or symbol may have completely different meanings for different individuals. Black is the color that the Western world associates with death and mourning. For other cultures it may be white or red. Our culture teaches us what responses are acceptable and appropriate, but even within the same culture a particular event may bring widely varied responses. It happens all the time between parents and children. The youth who celebrates a newly acquired driver's license as a symbol of his independence and maturity knows that his parents may view it with apprehension and misgivings. It's just that in flying we can make this alteration of perception easily and within ourselves.

We can use our understanding of how individuals can put different meanings on the same symbol as a bridge to understanding how we can do the same thing within ourselves. At the risk of overextending the image, we can think of the bridge as being on the path to the different reality we live in when we are flying.

Let me tell a story that may help to illuminate this idea. I've told it many times, and when I do some people react to it as something bizarre, a strange combination of the unusual and the macabre. It was none of these. Each of us who took part has his own personal meaning of what happened that night.

I had just received my commercial pilot temporary certificate and had signed up for a course to add the instrument rating. Smilin' Jack, the instructor, and I had just finished an instrument lesson, and the sun was going down as we went back to the hangar. While we were tying down the airplane, he asked me casually if I wanted a chance to use my new commercial certificate and make a "little money." Of course I accepted, and as Smilin' Jack began to explain the details I noticed three men standing in the shadows just inside the open hangar door. Each was dressed in a dark suit, each was in his early twenties, and each stood very stiffly. The man in the center, whose stance couldn't mask his anxiety, held a container that looked like a coffee can. It was wrapped in brown burlap and tied with a white ribbon. As we drew closer to the hangar, Smilin' Jack continuing his explanation in a quieter voice with each step, the situation became clear. My first job as a commercial pilot was to be a funeral flight over the Pacific Ocean. A young woman had committed suicide, and her husband, the distraught young man with the container, wanted to scatter her ashes over the ocean. His friends were here to accompany him.

Smilin' Jack greeted the three men. He introduced me to them as their pilot, and we all shook hands and tried to force polite smiles. Awkwardly we all walked through the hangar to the office. As the three men stood silently in the lobby, Smilin' Jack and I stepped into the tiny inner office.

"Take the 172. Head for Half Moon Bay and climb to about 3,000 feet. That'll get you over the hills on the peninsula. When you leave Hayward call Bay Control on 124.7 and ask them to keep you advised of traffic on the ILS to San Francisco. After you cross the shoreline fly outbound for another five or six minutes to put yourself over the drop area. Lower the flaps and put the airplane in slow flight to make it easy to open the window and keep the ashes from blowing back in. The guy who will sit in front with you is a pilot and wants to do some of the flying, so let him. I'll be gone when you get back so put the key in the mail drop. I'll pay you later."

I took the keys and the checklist and led the three men back out through the hangar to the airplane. They got in and got settled while I completed the line check. The man with the container sat in the left rear seat. When I finished I got in and started the engine, and called Ground Control for a taxi clearance. My passengers were still silent, and I felt as if I were intruding on their grief when I spoke to the controller.

We left Hayward, turned toward the bay, and began our climb. Again feeling that I was violating the solemnity of the moment, I called Bay Control for traffic advisories.

"Monitor San Francisco Approach on 126.8."

"Roger. Thank you."

Back to silence in the airplane. It is almost impossible to carry on a conversation in a small airplane that is climbing with full power, but the atmosphere in the cabin had become soundless. There was little to communicate between us anyhow. They were wrapped in thoughts of their own, and I was wrapped in thoughts of mine. I was a professional pilot for the first time. I had a certificate that said I was good enough to be trusted to fly with passengers and be paid for it. Now I was proving to the world and myself that it was true—I really was good enough.

I leveled off at 3,000 feet and turned to head directly toward the ocean. A full moon was setting in the southwest, and the sky was clear except for clouds that obscured the horizon far out to sea. It was nearly dark.

I turned to the man sitting to my right and broke the silence.

"You want to take it for a while? 210° at 3,000 feet."

"Okay."

He took the wheel, and I put my hands on my lap and pulled my feet off the pedals. I resented slightly having to give up control on my first working flight, but I reminded myself that I was still pilot in command and responsible for everything that happened. The red light on the ceiling behind my head was on, and it illuminated the control panel. I watched the instruments and could tell he was flying well, and I was relieved.

When we crossed the shoreline, I took back the controls. He had had enough time to make his symbolic gesture and fulfill his obligation to his friend.

Three minutes past the shoreline, over the sea. Suddenly there was a flickering light in the rear seat. My heart sank, then rose into my throat. A fire? Six miles from land?

No. My two rear passengers had lit a candle, and were reading something aloud by its light. The pilot in command made a sudden change of plans.

"Okay. We're here."

I quickly eased the airplane up into the slow flight attitude. It was time to get this over with and go home. Carburetor heat on. Throttle back. Flaps down. Throttle advanced. Trim for straight and level flight. There was no visible horizon, so I had to rely on the instruments to keep the airplane level and to maintain altitude. The candle was out now. I opened the window at my left.

"Okay."

"The moon! Turn so I can do it in the light of the moon!"

I turned to the right and lifted the left wing, keeping the airplane in a level slip to hold the moonlight.

"Okay?"

"Would you move forward? I can't reach the window."

I leaned forward and so did he, pushing the back of my seat forward as well. I scanned the instruments, but I couldn't see them. When my seat was pushed forward, it came between the instrument panel and the red light in the ceiling. It was dark, there was no horizon, and I couldn't see the instruments. The cabin was flooded with moonlight, but none fell on the artificial horizon in the upper left corner of the instrument panel.

I could feel that the airplane was in a descent and was building up airspeed, but I didn't know how much of either. I brought the wheel back to stop the dive, trying to reduce the airspeed at the same time. By now my chest was pressed against the control wheel, and at gross weight with a 30° bank I worried about stalling.

"Finished?"

"Not yet!"

At last I managed to see the needle on the altimeter. We were at 2,500 feet. I knew that if I could keep the altimeter steady and the moon on our left side, everything would be all right. I could sense some movement in the rear seat, but I didn't dare turn to see what was going on. Besides, I was sandwiched between the control wheel and the back of my seat. The movement seemed to stop.

"Are you finished?"

"Yes. I'm finished."

The ride home was silent, but silent in a different way than before. The ordeal was over and now there was only numbness. The lights of the San Mateo Bridge led almost directly to the Hayward Airport, and we made an easy, graceful landing.

The men each shook hands and thanked me as I was tying down the airplane. I tried to mumble something appropriate, and then they walked away in the dark to the parking lot. Our flight was over.

Smilin' Jack had forgotten about our arrangement when I reminded him the next week. He paid me twelve dollars.

I never saw the grieving widower and his friends again. Hopefully time has softened the anguish of that terrible event and their memories of it are now dim and indistinct. My memory of the flight is as clear as if it happened last night. For me it was an episode that marked an expectant starting point, not the painful end to a chapter of life that it was to the others.

Let's return from the past to the present since it is only the present we are concerned with in flight. At this moment we need to examine how flying can change our perceptions of familiar and important elements of our lives. Again, understanding this process is a crucial step in understanding how flying can restructure our notions of ourselves and the universe.

Below us is the ordinary world in which we live. The cities, the roads, the hills, the rivers are all part of our daily lives and routines. There's our house, there's the shopping center, there's the gas station where we stopped on the way to the airport. Ahead and to the left is the farm where we once took the children to buy a pumpkin for Halloween. All these places are still there, but their meaning to us is not the same. Now they have become just places beneath us, important only in relation to our flight. No matter how much of our lives we have invested in these places, the symbolic meaning they have for us in our ordinary existence is now gone. It will return for us when we return to the earth and to our ordinary life, but it is gone for now.

A few miles from where I now live and work, across the bay, is the school where I taught for fifteen years and the city where I lived for ten. I often fly in that area, departing or returning from a cross-country flight or using some of the remaining agricultural sections to practice ground reference maneuvers with students. What is beneath me is the stage on which many of the most important roles of my life were played. Whatever I am now was shaped in part by the events of those years in those places. The range of experiences was broad—birth, death; love, hate; joy, fear—encompassing much of what can happen in a person's life. I seldom go to that area on the ground, but when I do, each place I visit that was the scene of even a trace of these experiences triggers an echo of the emotions I felt there. The neighbors I knew have long since moved, but I still have friends at the school. Some are very close friends, and I always enjoy seeing them. But I can't bring myself to visit them at the school because I know that the setting would revive a barrage of emotions that I don't want to contend with and that would only interfere with the pleasures of friendship.

But none of this exists for me when I fly over these same places. They become landmarks, perhaps, if appropriate to a particular flight. Their symbolic meaning has been drained away, and they have no more significance than a road sign has to a traveler on a familiar highway.

If flying compels us to reinterpret familiar elements of our lives, it also demands that we reconsider our ideas about ourselves and our capabilities. Most of us, for whatever reasons, depend on others with specialized skills and knowledge to provide important services that we can't provide for ourselves. We depend on others in this manner in many significant ways; in fact the specialized aspect of labor in our society has been steadily growing stronger and more deeply established. We may pursue do-it-yourself projects as a hobby, but if we want a proper job done, we call in an expert. None of us is expected to be particularly competent outside of the narrow range of our own specialization.

The airplane we are flying in was built on an assembly line. It was put together by a collection of workers, each with his own particular responsibility for doing his assigned task to specific standards of quality. The person who rigged the control cables obviously did a good job or else we wouldn't be here, but that person had nothing to do with the installation of the engine, for example. That person couldn't do his job until those responsible for the assembly of the fuselage had finished with their work. The engine itself was manufactured elsewhere by a similar process of division of labor and then shipped to the aircraft factory. Someone else had to see that the engines were ordered and delivered in time to be installed in accordance with the production schedule. Before production started the airplane was designed by aeronautical engineers, marketing studies were conducted, financing arranged, and assembly workers hired and trained.

This is a typical production method in our society and reflects the rational, reductionist mode of thinking that is so pervasive. The process is analyzed and divided into specific discrete parts, and workers are hired to perform limited but tightly defined operations. The individual worker becomes a specialist in his particular operation, and he needs only enough knowledge of what his fellow workers are doing to ensure a smooth and orderly flow of work. They, of course, need only this same knowledge of what he does. What he does is not their responsibility. Their own responsibilities are clearly defined in a series of job descriptions that tell them precisely what they must do.

The similarity to other areas of our society is obvious. Bureaucracies, private or public, are based on this same model. When I was a teacher, I seldom felt compelled to enlighten students in my English or history classes about the basic principles of science or math. If the students were deficient in these subjects, I just complained about the poor job my colleagues were doing. If I did have to include materials on science and math, I tempted the wrath of my colleagues for intruding in their assigned domains, as well as taking the chance that I might not complete the curriculum I was required to teach. If students wrote poorly in their science papers, their teachers naturally carped at me. Later, when I wrote training programs for a private consulting firm, I had to be sure that each program had a reasonable degree of internal consistency. As long as what I produced was logical and made sense on paper, I had done my job properly. It didn't matter that I knew nothing about repairing electric motors or welding pipe, the subjects the programs were supposed to teach. After all, I was a senior training analyst, and it was up to someone else to try to teach from the programs I wrote. If the programs failed, it wasn't my fault, nor was it the fault of the instructors since they couldn't be expected to work from materials they hadn't prepared themselves.

When flying an airplane, this habit of limited responsibility and easy shifting of blame comes to an end. The airplane may have been built by a collection of specialists, but now the pilot must accept complete responsibility for the flight. In a unique and total way he becomes fully responsible for his own life, and the lives of his passengers. He can only depend on himself. It is his skill, his knowledge, his perceptions, his judgments that he must depend on. There is no one else to turn to. There are no underlings to upbraid, no superiors to conspire against, no fellow workers to silently criticize and feel superior to. Someone must bring the airplane safely to the ground at the end of the flight, and that is the pilot's job. It's too late to call in an airport approach specialist or a landing expert. He must be these and more himself. All responsibility rests with him.

Where else can one become, if even for a short time, so completely responsible for one's own life? There is a paradox here because man cannot live in total isolation from others. He is a social animal and can exist only by interacting with others. Even the hermit must have a human society from which to escape. But the pilot is temporarily cut off from others. His culture has shaped him and has given direction to his life, and now only he can ensure that his life will continue. He can only turn to himself. How rare this is in our society, or in any other. Most of us have lives of complex interdependency, relying on others to respond to the social roles we choose to play or are obliged to assume. Without others to respond to us, our roles are invalid, so we have to depend on others for self-definition and all that may flow from it. For the time the pilot is in the air, he has set aside his dependence on others, and his life is truly his own.

This is powerful knowledge, to know that we can assume full responsibility for our own lives. There is no choice, really; if we fly, we must accept it. Most of us never have the chance to take control of our lives this way and can never have the experience of being so self-reliant. In our ordinary lives, on the ground, there are precious few opportunities to do so, and most of us can hardly conceive of the possibility. Our society and the reality it has created simply doesn't allow for it. It's not thought to be possible, so it doesn't exist.

Since this supreme self-sufficiency is not thought to exist, there are no guidelines or directions we can seek to try to find it in our normal lives on the ground. Each human culture has its own unique way of defining the nature of the universe and what is contained within it; this agreement upon what is real and what is possible is one of the important elements that holds a society together. Certain things must be accepted as given so that society can get on with the task of survival, and survival will be much more likely if each member perceives the universe through a common frame of reference. One of the functions of human culture is to act as a perceptual screening device, filtering the infinite possibilities of the universe into a manageable and agreed-on structure. When we are born into a culture, we come to accept its description of reality as true. We have no choice, really, since no other model is presented for comparison. What we must realize is that any culture's description of reality is imperfect since it is designed to provide a sense of social continuity by reinforcing established traditions and values. It is by necessity incomplete, and there is no way for us to understand this unless we can somehow step outside of the reality we have learned to accept.

Flying gives us the opportunity to do this, if we are willing to take the step. Since we are assuming that reality is a rather arbitrary construct that society forges for convenience and common reference, let's try to illustrate how this works by looking at some examples of the symbolic representations of reality.

Just as you and I need a vision of reality to guide us through life, the motorist needs a road map to guide him to his destination. It's one way of representing a geographic area, and it shows towns, cities, and other obvious landmarks that will help the motorist find his way through unfamiliar places. The road map shows features of the territory relevant to the motorist, such as street names and numbers, since they are important symbols for determining where he is. He understands that the symbols on the map are not complete and that there are more things in a given area than could possibly be included on his road map. But he accepts this because he knows that it isn't important so long as the map serves its purpose, and he can depend on it to help him find his way.

The pilot's navigational chart is also incomplete, but it includes much that is not shown on the motorist's road map. The pilot needs to know the heights of hills and mountains so that he can choose a safe altitude to fly over them. He needs to know the location and frequencies of radio beacons and the configurations of lakes, rivers, and other easily recognized landmarks to establish where he is. All this information is symbolized for him on his chart. The chart depicts the same territory as the road map, but not in the same way. Each is a different reflection of the territory, and they converge only when elements of the territory converge by necessity. Both the motorist and the pilot need to know, for example, the locations of towns and freeways since they are important to each, although for different reasons. The central point is that neither depiction is complete, and each contains elements that the other omits.

This is simple and easy to accept; yet most of us fall into a trap when we apply the same idea to our own lives and to the world in which we live. The trap is an extension of the semantic one that S. I. Hayakawa described years ago concerning the relationship between language and reality. He explained that a word (the map, in our case) is a symbol for a thing, not the thing itself (the territory). The symbol is meant simply to be a representation of something else. It stands for something but is not the thing itself.* A society's construction of reality, its description of the universe, is analogous to a map. It tells us, in general, what we need to know to get along, but like the map, it is incomplete. We tend to mistake our map of reality for the universe itself, denying the existence or even the possibility of the existence of anything not symbolized on it. We forget, or never knew, that our reality was created for a specific social purpose and that it cannot be a complete representation of all that could be.

It is only when we confuse what we accept as reality for the universe that we deny ourselves the discovery of other constellations of possibilities. Most of us are familiar with the reality of the motorist. That is the ordinary reality we have in common. The pilot can take part in this reality since it is his as well. But he knows that it is not the only reality, that it is but one possible arrangement for perceiving what the universe might contain. Flying into the air above the inversion layer was a

---

* S. I. Hayakawa, *Language in Thought and Action* (New York: Harcourt, Brace & World, 1949), pp. 31–33.

clue. When we live all our life on the ground, we are likely to accept smog and pollution as intrinsic parts of the universe, givens. They blanket our existence, and it becomes difficult to imagine a world without them—until we enter the sky above the pollution and are compelled to realize that other alternatives are possible. But we had to leave our ordinary environment to understand this. The wag who said, "No one knows who discovered water, but it certainly wasn't a fish," was right.

The experience of flying gives the pilot the chance to move out of his ordinary reality, and all this has become apparent to him. He has had to reorder his world while he flies, and he has learned that it is possible to do so. We can restructure reality and not go mad, and still survive. We can create our own world. This is what the pilot does. While he is in the air he is in a different reality, a world different from the one in which he exists when he is on the ground.

Now that we've been in the air in straight and level flight, we can begin to understand that flying is a way that the pilot can alter his relationship to the earth, to society, and even to his own past. Little by little he discovers that he is part of a reality that doesn't prevail in his life on the ground. He has altered his notion of the universe and what might exist and take place in it.

In a very loose sense we have been considering what flying may be, but we must be careful to remember that the tools we have been using to describe flying have been words, and as we just discussed words are symbols of reality, not reality itself. Our difficulties have been compounded by the fact that we are trying to describe aspects of flying that don't have parallels in our ordinary existence and so don't have words commonly understood to stand for these elements of the pilot's world. The usual language of flying won't do either since we aren't concerned here with ordinary descriptions and terminology. Our job would be easier if we could use aviation terms; if we came upon a word or idea that was unfamiliar, we could consult one of many recognized authorities for an acceptable, agreed-on definition. In technical language the symbol and the reality tend to be well matched. Several times we have mentioned things that seem paradoxical and inconsistent. In terms of language and how it is generally used, they haven't made sense because they tended to violate the logic of language itself. Our language wasn't designed to be used as we are using it. It was intended to reflect ordinary reality, and here we have been trying to use it to reach into something else.

In this case the best solution seems to be to make the words as clear as we can by giving them a context, a setting in which they can be of the greatest possible significance to us. That setting, of course, is our own experience. We have to interpret the words and their approximate meanings, testing and measuring, shaping and fitting into what we know to be true about our own lives. It doesn't matter if others don't agree. We know what is true about our own lives, just as the pilot knows that he experiences the world in a different way from his fellows who never leave the ground.

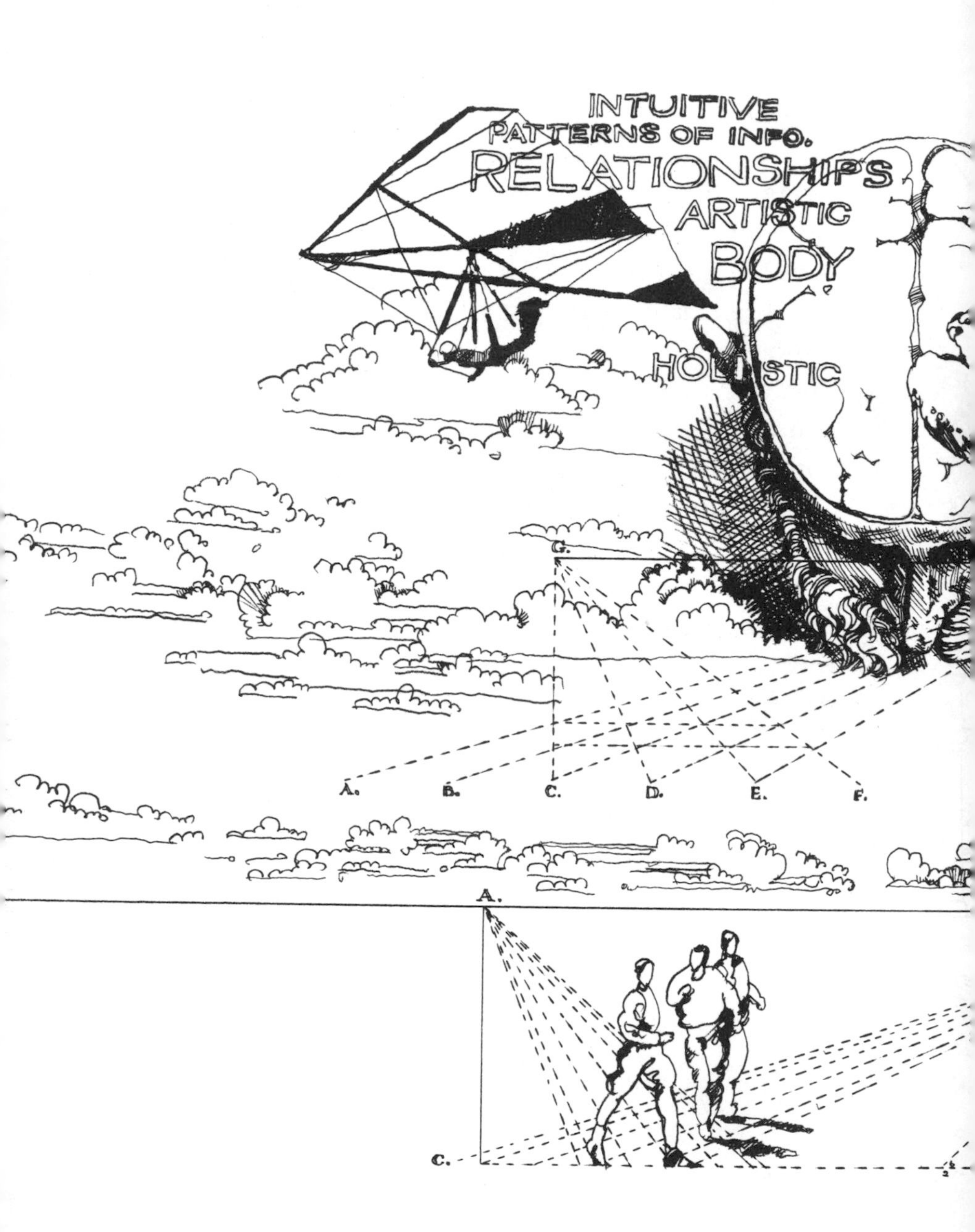
INTUITIVE
PATTERNS OF INFO.
RELATIONSHIPS
ARTISTIC
BODY
HOLISTIC
G.
A.
B.
C.
D.
E.
F.
A.
C.

LOGICAL
VERBAL
ANALYTIC
STEP BY STEP
PROGRESSION
STATISTICS
MIND
PROGRESSION
H.
B.

3. The Turn

Perceiving the Turn

THE turn is the most complex of flying's four basic maneuvers. The pilot must apply coordinated pressures to all the control surfaces so that the airplane will turn at a constant rate, at a constant bank, and at a constant altitude. It's a fine example of classical physics, of Newtonian action and reaction forces in harmony with each other. It's also a symbol of how flying involves a combination of perceptual modes and ways of thinking that we ordinarily think of as diverse and unrelated.

Once the turn is started, the first thing we are aware of is that our wings are no longer parallel to the horizon. If we look out the window in the direction of the turn, we can look down and see the ground. If we look out the opposite window, we can only see the sky. The airplane has rolled around its longitudinal axis, an imaginary line that runs from the nose of the airplane to the tail. The rolling is caused by the ailerons, the control surfaces on the trailing edges of the wing near the tips. They work in opposition to each other so that when one goes up the other goes down. The aileron that goes up reduces the amount of lift being produced by the wing on that side and the wing lowers, at the same time the other aileron goes down, increasing the amount of lift being produced on its side, and that wing rises. When the proper degree of bank is reached the pilot neutralizes the controls by aligning the ailerons with the wings again, and the airplane stabilizes in a bank at that angle to the horizon.

It was the Wright brothers who developed the idea of banking an airplane in order to make it turn. Most of the others who were building flying machines around the turn of the century were trying to find a way to make a turn and still allow the wings to remain parallel to the ground—something like the way one turns a slow-moving carriage, horseless or otherwise. Instead of the movable controls on the wings used now, the Wrights rigged their early Fliers so that the trailing edges of the wings could be twisted slightly to achieve the same effect. After their success at Kitty Hawk they spent several years in Ohio perfecting their flying technique while they continued improving the general design and performance of their airplane.

The turning system having proved to be the key to controllable flight, they realized the importance of establishing patent rights to their airplane. Their fear of someone copying their ideas led to an obsession with secrecy that bordered on paranoia. Outside of Dayton they rented a practice field where they stored their machine in a shed that also served as their workshop. A trolley line ran nearby, but a line of trees blocked the view of the field from anyone on the train. Just to make sure that no passenger might catch a glimpse, they timed their flights so that the aircraft was always in the shed or on the ground when the trolley was due to go by.

Their work had reduced the problem of controlling an airplane to its basic principles, and has not been significantly improved to this day. As in all good design the beauty of their plan lay in its simplicity.

But more than the ailerons are involved in the turn. When the wing is banked, the force of lift is tilted out of alignment with the force of gravity. Since some of the lift acting to counteract gravity is lost, something must be done to increase the amount of lift the wing produces so that a resultant force equal to the force of gravity keeps the airplane at a constant altitude while it is turning. To do this requires a slight amount of increase of pitch, or motion around the airplane's lateral axis—another imaginary line that runs from wingtip to wingtip and crosses the longitudinal axis. By bringing the nose up slightly the additional lift is produced and a constant altitude is maintained; however, the increased pitch attitude causes a reduction in airspeed.

One more control is left to add to this equation: the rudder, or vertical stabilizer. The pilot operates this with his feet, and it acts much like the rudder on a boat. Step on the right pedal and the nose of the airplane will swing to the right; step on the left pedal and the nose will swing to the left. This motion is called "yaw" and turns the airplane around its vertical axis: another imaginary line that intersects the other two axes at right angles to them.

The correct amount of rudder pressure keeps the longitudinal axis tangent to the circle of the turn and keeps the centrifugal force, which acts to move the airplane to the outside of the turn, and the centripetal force—its equal and opposite counterpart—in balance with each other. In reality the rudder doesn't turn the airplane; the ailerons that cause the wing to bank do that. The rudder compensates for the forces of adverse yaw, a tendency to turn in the opposite direction of the bank, and holds the airplane and all the forces the turn generates in equilibrium.

This description and analysis of the turn could be extended and broken down further into more detail. Air density, wing loading, and asymmetrical thrust are just a few of the additional factors that influence the airplane's performance in a turn and that would become clear to us if we continued to examine what the turn is in this rational, analytical manner. Pages of graphs, diagrams, and tables could be developed to abstract these elements into symbolic forms that would help us understand the turn in a logical and scientific way. This way of understanding is valid and important. It reflects the capacity of the left hemisphere of our brain to process information in a step-by-step, sequential procedure, but it is not the only way we have of experiencing what a turn may be, and it is not the only way a pilot can experience a turn.

The pilot must also be able to experience the turn as movement through space. He has to sense the path his airplane is taking through the sky in relation to what he can see and feel through his senses. The relationship of the airplane's nose to the horizon will tell him if he is holding his altitude and a constant bank angle. If he imagines that his airplane has a giant paintbrush extending from its nose, he can watch it "draw" a straight line across the sky parallel to the horizon—if he maintains constant altitude in the turn. When the line begins to waver, to curve up or down, he knows that he must adjust the control pressures to make the line straight again. When the line he paints is straight, his altitude is constant. The way his body feels will tell him if the forces in the turn are balanced with the components of lift, weight, and gravity. When the airplane is in a bank of 60°, a force equal to twice that of gravity is acting on him, and his body will seem to feel twice as heavy as it does on earth. This is called "pulling G's," and it occurs in every turn in proportion to the angle of the bank. The experienced pilot can detect these forces acting on his body even when their magnitude is very slight.

The Pilot's Ways of Knowing the World

BOTH these ways of perceiving the turn are valid. But both are incomplete, just as the road map and aviation chart are incomplete reflections of reality. But they are complementary since they depict different aspects of the same thing. In terms of brain functions it's the right hemisphere—the side that deals best with spatial relationships and the integration of data in patterns—that guides the pilot in this maneuver. As we discussed before, he has learned to bypass his rational instincts and to act without hesitation on what he perceives, without analysis and abstraction. He becomes part of the environment: the airplane, the atmosphere, the horizon—responding as part of a whole, keeping all the elements in a state of dynamic balance. With this in mind we can now examine how the two sides of the brain have, in addition to specialized modes of perception, specialized modes of consciousness as well.

All of us are capable of experiencing more than our normal mode of consciousness, our ordinary reality. To be able to do so is part of being human since we come equipped with the biological apparatus to make it possible. Why we don't cultivate an appreciation of this fact and develop our capacity to do so is mainly a reflection of the bias of our culture. The rational, left hemisphere mode of perception and our familiar mode of consciousness analogous to it have been paramount in Western civilization for so long that any other style has been overshadowed or denied. Given the achievements of the scientific approach, it's easy to understand why this could happen. Who wants to argue with success? But on the other hand, it doesn't make sense to deny part of our potential as human beings. In an often-quoted statement, Robert Ornstein gives us a clue to how we could have ignored this aspect of our being:

Suppose there were two completely independent groups of investigators; one group (scientists) works exclusively during the day, the other (the mysterious esoteric psychologists) works exclusively in the night, neither communicating well with the other. If those who work at night look up and see the faint stars, they may produce documents which predict the positions of the stars at any given time, but these writings will be totally incomprehensible to someone who experiences only daylight. The brilliance of the sun obscures the subtle light of the stars from view.

A modern scientific psychologist may read in an obscure esoteric book about the existence of these "points of subtle light" and attempt to locate them, by his usual means of investigation—during the daytime. No matter how open and honest the investigator, and how dutiful his observations, he will be simply unable to find the subtle light in the brilliance of the day. The failure of many scientific investigators to locate the "subtle light" will only strengthen their conviction that "stars" do not exist. The methods of science, then, have largely focused on one mode of knowing—identified with daytime. Contemporary science has developed methodology and discovered laws which are valid and which have proven essential to the development of our civilization, but which may be only a special case, as Newtonian mechanics is to Einsteinian.*

---

* Robert E. Ornstein, *The Psychology of Consciousness,* Second Edition (New York: Harcourt Brace Jovanovich, Inc., 1977), pp. 12–13.

The notion that there is a complementary duality of consciousness is of course not new, even if we seem to be discussing it here as if it were a contemporary revelation and not something that our culture has tended to neglect. The two modes of consciousness are analogous to the two modes of perception for which each side of the brain has its own specialization. Numerous metaphors from many sources attempt to illuminate this dual aspect of our nature. Here are just a few, adapted from Ornstein:*

DUAL MODES OF PERCEPTION

PROPOSED BY	RATIONAL	INTUITIVE
Many sources	Left hemisphere	Right hemisphere
Many sources	Day	Night
Blackburn	Intellectual	Sensuous
Deikman	Active	Receptive
Levy, Sperry	Analytic	Gestalt
Lee	Lineal	Nonlineal
Luria	Sequential	Simultaneous
I Ching	The Creative: heaven masculine, Yang	The Receptive: earth feminine, Yin
I Ching	Light	Dark
I Ching	Time	Space
Many sources	Verbal	Spatial
Many sources	Intellectual	Intuitive
Bacon	Argument	Experience

* *Ibid.*, p. 37.

We can see that in many cultural, philosophical, and scientific traditions there is a recognition that man is not limited in the ways he can deal with his surroundings. He is able to perceive and experience his world in at least two modes, as these metaphors suggest, and the balance between the two extremes is left to us to determine.

To the pilot's joy, he can exist in more than one mode of consciousness. He is not limited to the ordinary, rational mode of everyday life, although he can function successfully within it. The right hemisphere is there for him too, and he can combine the two as he extends the boundaries of human possibilities. The greatest figures in the history of aviation have done just this.

Wilbur Wright combined the qualities of a brilliant scientist with those of a creative genius. Wilbur, the older of the two brothers, seems to have made the larger contribution to their joint achievement. He not only invented the airplane; he learned to fly it as well. Certainly he relied heavily on the work of earlier inventors who tried to find a way to fly, but Wilbur was the one who put the ideas and insights together in a simple and successful way. He sensed the relationships between the forces acting on an aircraft and created a control system that would respond to those forces. Intuitive thinking seems to be the hallmark of all successful inventors.

But the creation of ideas is only part of the process. The ideas themselves must be rigorously tested and refined to be sure they are truly valid and workable. Again, Wilbur showed himself to be adept at this chore as well. It is this testing of a theory, the rational process of acquiring knowledge through experimentation, that is such an important part of the work of the scientist. Fabricating a wind tunnel in the Wright bicycle shop in Dayton, Wilbur tested and collected data on 150 model airfoils—wing shapes of various dimensions. The wind tunnel idea had been used before for this purpose, but it was Wilbur who, in 1901, designed the instrumentation necessary to accurately measure and evaluate the data that the wind tunnel was able to generate. It is obvious that he had a firm grasp of the complex mathematics involved in this kind of study. From the wind tunnel experiments he determined which airfoils could be expected to be the most effective when incorporated into the design of a full-sized aircraft. He was not dissuaded by the fact that his findings ran counter to the pronouncements of the accepted scientific authorities of the day. What he learned from his wind tunnel experiments was put to practical application in the successful "Wright Flier" of 1903, neatly demonstrating the relationship of the rational and intuitive modes of consciousness and thought. That Wilbur Wright sensed the dual levels of consciousness possible for a pilot to experience is clear. He described the situation succinctly in 1906, after he and his brother had each flown many times: "When you know, after the first few minutes, that the whole mechanism is working perfectly, the sensation is so keenly delightful as to be almost beyond description. More than anything else the sensation is one of perfect peace, mingled with an excitement that strains every nerve to the utmost, if you can conceive of such a combination."*

* John Evangelist Walsh, *One Day at Kitty Hawk* (New York: Thomas Y. Crowell, 1975), p. 188.

The most celebrated of American aviation pioneers, Charles Lindbergh, also symbolized the full use and appreciation of man's duality of mind. Behind the hero and enigmatic public figure we find a complex and remarkably gifted man. He managed to combine in his life the detached, objective intelligence of the scientist with the all-encompassing point of view of a mystic who felt himself part of a seamless web of existence. Lindbergh's flight to Paris and his contributions to aviation are part of the common lore of our culture and don't need to be recounted here. What is less well known is the work he did in early rocket research with Robert Goddard and other experimenters, and his work with Alexis Carrel in the development of artificial organs that could be implanted in humans.

Lindbergh's reputation as a rational man of science is firmly established, but he was obviously aware of another dimension of consciousness that surely exists but is not verifiable by the methods of science. In *The Spirit of St. Louis*, his story of the Paris flight, he tells of the phantoms who rode with him for the last few hours over the Atlantic Ocean. "Friendly, vapor-like shapes, without substance, yet real," he accepted their presence and their existence as a part of the universe of which we are not normally aware. "There's no suddenness to their appearance. Without turning my head, I see them as clearly as though in my normal field of vision. There's no limit to my sight—my skull is one great eye, seeing everywhere at once. . . . At times, voices come out of the air itself, clear yet far away, traveling through distances that can't be measured by the scale of human miles; familiar voices, conversing and advising on my flight, discussing problems of my navigation, reassuring me, giving me messages of importance unattainable in ordinary life."*

As a result of this and other experiences he developed an attitude toward existence that left open and unlimited man's ultimate capacity for experience and knowledge. Even the line between life and death is to be questioned since if man is part of the universe, Lindbergh speculated, why should his existence be confined to that instant of time when the collection of molecules that make up his physical being is charged with what is called "life"? Lindbergh saw death as a part of a natural cycle, not as an end but as the beginning of a different form of existence that serves to demonstrate man's interrelationship with all that is in the Universe. At the end of *Autobiography of Values*, published after his death in 1974, he wrote: "My aging body transmits an ageless life stream. Molecular and atomic replacement change life's composition. Molecules take part in structure and in training, countless trillions of them. After my death, the molecules of my being will return to the earth and the sky. They came from the stars. I am of the stars."§

So, in a number of ways, the pilot has the opportunity to open himself to a mode of consciousness different from, yet complementary to, the rational style that is dominant in our culture. Since more than a single, culturally determined means of perceiving reality is available to us, indeed is part of our nature as human beings, there is a danger in applying arbitrary

---

* Charles A. Lindbergh, *The Spirit of St. Louis* (New York: Charles Scribner's Sons, 1953), p. 389.

§ Charles A. Lindbergh, *Autobiography of Values* (New York: Harcourt Brace Jovanovich, 1978), p. 402.

limits on ourselves and not exploring all that we might be. Some sort of balance is necessary. The domination of rational thought in Western civilization has produced spectacular human achievements and has enriched our lives, but overdependence on it has its risks, as all of us who live in the nuclear age know only too well. Lindbergh, this time at a bomber factory at Willow Run during World War II, said: "This was a temple of the god of science at which we moderns worshipped. Here was the efficiency, the superhuman magic of which we had dreamed. Only two years before on this same spot, I would have been surrounded by hickories, maples, and oaks. Scientific man could now touch a forest in Michigan with his wand, and by so doing wipe out European cities. . . . How could we further human progress by striving for such scientific goals when the very concentration on them blinded us to higher values, mocked the brotherhood of man, shielded us from God?"*

When viewed from the full span of human history, of course, scientific thought and the experimental method of acquiring knowledge are relatively recent developments. The traditions of Eastern thought stretch back much farther in time. Rather than stressing the separateness of things in the universe—man and society, mind and body, spirit and matter—the East tends to perceive the universe as a unified whole. The idea that man can somehow stand outside the physical world while he measures, classifies, and manipulates its parts is incomprehensible in Eastern philosophy; yet this has been a basic assumption of

* Charles A. Lindbergh, "Of Flight and Life," in *The Saga of Flight*, ed. Neville Duke and Edward Lanchberg (New York: The John Day Company, 1961), p. 199.

Western thought. So has been the assumption that all things can be broken down and categorized into irreducible, discrete parts. This is the model of modern bureaucracy, the Dewey decimal system, and the techniques of mass production, all of which lie outside the realm of Eastern traditions, which hold that all is interdependent and interrelated and that nothing can exist in isolation. Each school of thought represents a kind of specialization, each with its own areas of appropriate use, but neither capable of universal application. Again, we seem to be back to our image of the road maps and aviation charts, both useful to a certain extent, but both having their own intrinsic limitations.

The possibility for combining these two approaches to reality can occur in unlikely places. In an extremely provocative book, *The Tao of Physics*, Fritjof Capra shows how a synthesis of Eastern and Western philosophies exists in one of the most sophisticated areas of modern science, subatomic physics. In the world of subatomic particles, the observer can't stand apart from what is being observed since the act of observation itself significantly influences whatever phenomenon is taking place. Even the notion of a particle has its limitations since light, for example, can be considered both as a collection of particles and as wave motion, depending upon how the light is related to other factors under study. Capra points out that atomic particles cannot be defined and understood as independent entities but only as interrelated parts of a larger whole. The laws of classical Newtonian physics, which seem to be applied universally in everyday life, have long since been found inapplicable to the inside of the atom. The subatomic world lies outside our ordinary sensory perceptions and can be perceived only indirectly, through the use of precise scientific instruments that allow the physicist to see the results of a phenomenon—such as the collision of atomic particles in a cloud chamber—not the phenomenon itself. When he attempts to analyze the tracks that the particles have made in the chamber, he finds that the properties of the particles that he observes are dependent on both the manner in which the particles have been prepared for the experiment and the manner in which the measurement has been taken. As the preparation and measurement change, so do the properties of the particles. The physicist also has great difficulty translating and expressing into ordinary language what he has learned about the atom. Just as the Zen notion of the sound of one hand clapping doesn't easily lend itself to words, neither do concepts such as relativity, curved space, and elastic time. Capra maintains that the physicist, in a number of ways, may achieve a complementary balance of unified relationships and specific data, just as the essence of Tao is a dynamic balance of Yin and Yang.

Our airplane is keeping its dynamic balance with its environment, turning circles in the sky. We are part of that environment, responding to the slightest shift away from equilibrium, matching complementary modes of perception and consciousness, coordinating mind and body to hold the airplane in the turn.

The Worlds Within Us

WHAT are we doing here, anyhow? Would any sane person fly around in the air in an aluminum contraption propelled by an engine that has less horsepower than the average American automobile? Sure, we get a nice view of the countryside, but wouldn't a picture do just as well? Aren't there mountaintops from which we can see for miles without ever leaving the ground?

Of course there are, and there are Vista Points and Space Needles and World Trade Centers too, where we can see out to the horizon. But this flight has been one where we are not only looking out but also looking in—looking inward at ourselves to see what might be there that perhaps we didn't know about before—out past our inner horizons. And we have found some things, for sure. Flying isn't the only means to find what is beyond that horizon; there are more ways than we could imagine, but it is a way that will work for some of us. Because flying can be such a dramatic experience, its lessons can be sharp and vivid. After all we really do leave the world for a while.

We must remember that the complex technological society

we live in is a very recent development in human history. Man has been wandering about the planet for millions of years, in some form or other. Even agriculture, the systematic planting and harvesting of a food supply rather than dependence on what could be acquired by hunting and gathering, was developed less than 10,000 years ago. Without the reliable food source that agriculture provides and the binding of man to specific locations that it demands, villages, cities, and eventually highly organized civilizations probably would not have been possible. Man's biological evolution, however, has not necessarily molded him for life in the city. The survival skills we need in our society are not the same as those we would need if we lived by hunting and gathering, although we are physically suited for such an existence. That's how man has spent most of his life.

Flying simply demands that we put to active use senses and abilities that lie within us but that may have been overlooked or undervalued because they don't seem to be of great importance to survival in the twentieth century. Unless we are athletes, we seldom need to use our kinesthetic senses and our ability for coordinated movement of our bodies to the extent that we must when flying. All our other sensory systems must be at their peak in the air, not just our eyes and ears, but all our senses as we rediscover how we can use them all in combination. Our rational and intuitive capacities of perception are joined in flying since both are available to us and both must be put to use to make their particular contribution to our flight. The intense effort and the relaxation, the mental concentration and mental void, bring us to modes of consciousness that are so rare and precious that we think them exotic and associated with mysticism and forbidden pleasures.

But this isn't so. More than one mode of consciousness is possible because it is within our range of possibilities as human beings. When we put all this together in flying, it becomes clear that we are pushing back that inner horizon that defined the limits of what we are and could be. And we can begin to see that there is more to being human than we might have thought possible. Underneath the layers of cultural conditioning that obscure those aspects of life ignored or discounted in our modern industrial society, the definition of what it means to be human grows larger and greater.

Consider a major league outfielder waiting for the pitch to an opposing batter. Nothing exists for him at the moment except the game. His concentration is total; yet he is completely relaxed. He has positioned himself in the outfield very precisely because he is a careful observer, has studied the records, and knows just where the batter is likely to hit the ball. The pitch. The swing. The sound of the bat hitting the ball. Without stopping to analyze and reflect, the outfielder moves in the direction of the ball. With grace and agility he moves under the ball and captures it for the out. For a few seconds he is using a combination of skills and perceptions that the pilot must use the entire time that he is in the air. For the outfielder the stakes are an out, a game, or perhaps even the World Series. For the pilot, like the primitive hunter, the stakes are much higher. The prize is life itself.

At these times we are fully functioning human beings. All the capacities and potentials we evolved are being exercised, meshed, and integrated into a balanced whole. We are at the top of our form, and while we are extending the limits of human potential we are also getting a clearer understanding of just who we are as individuals. But beside all that, developing and using our full human potential are great fun. It brings exhilaration and happiness to be able to perform as a fulfilled and complete individual.

Flying seems to provide for some people the opportunity to become what psychologist Abraham Maslow called a "self-actualizing" person. Maslow turned the study of psychology upside-down by studying individuals he deemed healthy and well functioning, rather than those who suffered from some mental or emotional imbalance. A self-actualizing person is one who can bring all his powers together in an integrated, joyful, and efficient way, moving toward what he could potentially become. When a person does this, he makes his potentials real, actual, and thus becomes more truly human and more truly himself at the same time. Nothing is held back and no aspect of the person that can contribute to what he is doing is left idle and undirected.

Self-actualizing persons have many characteristics, some difficult to define and specify, others similar and overlapping and with only very subtle differences between them. But some basic characteristics stand out clearly and apply precisely to flying. For one, the self-actualizing person acts with a strong sense of purpose, with all his energies galvanized to the task at hand, a task that he himself has invested with meaning. There is no room for boredom or indifference since all awareness is focused on what he has chosen to be significant to him. All his capacities are brought to bear on it. In turn, since there is meaning in the task, there is meaning in life itself. For the pilot, whose life depends on pouring his entire being into his flight, life and his flight become synonymous.

But still this purpose lies outside, external to the self-actualizing person himself. It does not involve his sense of self since he knows himself and is not concerned with his own level of self-esteem. He has objective knowledge of who he is and what he can do. The self-actualizing person knows, as does the pilot, that he can and must accept responsibility for his actions. Although not immune to fear and anxiety, these feelings do not overwhelm and paralyze him. They are accepted for what they are and integrated into his notion of himself.

Earlier we discussed those times in flying when the rest of the world seems to go away, when we seem completely detached from ourselves, our lives, and even from our own ordinary consciousness—those times when we seem to be existing in an intense and dynamic relationship with all that is around us. Maslow called these times "peak experiences," "a spurt in which the powers of the person come together in a particularly efficient and intensely enjoyable way. . . ."* It is these moments that most vividly give us an idea of what we might be capable of, that give us a vision of what might be possible if we can generate the energy to open ourselves to what life might contain. Suddenly there *is* more, and for the moment, we are richer and life is more complete than we could ever imagine. We may have been accustomed to what we believed was a satisfactory existence, but now we are faced with something new. Our senses are keener. The world seems unified and whole, but we are more conscious than ever of the details of what is around us. The effect of these experiences is very powerful, and we

* Abraham H. Maslow, *Toward a Psychology of Being*, 2nd ed. (New York: Van Nostrand, 1968), p. 97.

probably couldn't tolerate living so intensely very often, but even these occasional peak experiences give us some clues to what might be within our reach.

Suppose that we had never seen the moon except when it was only quarter full. We were always satisfied with its beauty and content with its brilliance, unable to imagine anything brighter. Then one night, to our surprise, the moon is full. What we were not able to dream is real. We may not see a full moon again for some time, if ever, but now we have an idea of what is possible. In the words of one of Maslow's biographers, "For what is a peak experience but a glimpse of the full moon?"*

Experiences like these can happen in many different situations. Surely you can give your own examples. Once the personal chemistry is in balance, once we are investing ourselves in something as demanding and rewarding as flying, the experiences are apt to occur, and we are given a glimpse of what human nature and life can hold. Long before I learned to fly I learned about these moments, although I had no insight into their significance until years later. I think I have been lucky because, in some unconscious way and not because of any great perceptions on my part, I've known they could occur in the ordinary routine of life and could leave the outer boundaries of existence undefined and open-ended.

When I was in high school, I played in regional music festivals that were held each spring. Bands and orchestras were assembled from the best or the most ambitious players from a

---

* Colin Wilson, *New Pathways to Psychology* (New York: Taplinger, 1972), p. 244.

dozen or so schools. After auditions for principal chairs for each instrument, rehearsals were held on two successive Saturdays, with a concert on the evening of the second. The musicians were generally much better than I, but I enjoyed playing and being part of the group and didn't mind being in or near the last chair in my section.

The year the festival was held in my town, I was playing in the second flute section of the orchestra. We had been working hard all day, and later in the afternoon on the day of the concert, the conductor took us through a final rehearsal of Franck's D Minor Symphony. The conductor, Kurt Herbert Adler, was an experienced European professional who later had a long and successful career as general manager of the San Francisco Opera. I remember that we started well and seemed to improve as we continued. Then suddenly everything began to come together. The rhythm, the dynamics, the bowing, the fingering, the sheer passion of the music, everything blended as never before.

"Play! Play!" the conductor shouted at us, not in anger, but in joy and exhilaration.

When we finished, we sat back in quiet exhaustion. A group of adults listening at the back of the auditorium, my father among them, applauded spontaneously. The conductor tried to explain what had happened so that we might understand what he already knew. It didn't matter what might happen at the performance that night, he told us. Our experience that afternoon was enough and transcended anything that might occur later. We had touched the essence of the music and had touched on our own potentials as musicians. For just those few moments we had the chance to see how well we really could play and to see what we and the composer had within ourselves. The conductor's words—and mine as I write years later—could only allude to the experience but could not adequately describe it.

Let me tell another story, one not so formal perhaps, but one that reflects a similar kind of magnification of feeling that flowed naturally from the experience of flying. My first solo cross-country flight on which I was instructed to land at a particular field—and not just find any convenient place and return—was to a small country airport about fifty miles from my home field. The airport was easy to find since it was in a long valley, near a freeway and railroad tracks and only a few miles from the major town in that part of the valley. I found the airport on schedule, and after an abortive first try I landed safely. At that time student pilots had to have their flight logs signed at the destination of a cross-country flight—just to prove that they really got there—so I tied down the airplane and began to look for someone who might be official enough to provide a valid signature. There was a rusty Quonset hut a few hundred yards away, and I started walking toward it. It looked as if it might fall down at any moment.

As I approached I saw that the building had two doors, each with its own sign hanging above it. The closer sign said "Airport Office," the other said "Pilot Lounge." I started for the office door and hesitated. Wait a minute. Wasn't I a pilot? Didn't I fly here by myself, with no help from anyone? Of course I did. I'm a pilot and I'm entitled to enter the pilot lounge.

I was so swept up in the honest pride of achievement and the absurdity of the decision I was faced with that I began laughing out loud and floated the rest of the way to the door. "Oh, boy! This is like the first time I had a real ID card to prove I was old enough to buy a drink. I'm thirty-five years old, and now I can go into the pilot lounge on my own, just like the Real Pilots."

I opened the door. It was Monday morning and the lounge hadn't been cleared of its weekend accumulation of debris. Styrofoam cups were everywhere, several with insects drowned in the stale coffee. Ashtrays were festooned with broken cigarettes that spilled onto remnants of old aviation magazines. Messy, but who cares? Not me. This is the pilot lounge, and I can come here because I'm a pilot. Then I noticed that the doors I had seen from outside both opened into the same large room. The office and the pilot lounge were really the same place. I began to laugh even louder. A man with a bewildered look on his face entered from the rest room and agreed to sign my log. I left through the door marked "Pilot Lounge."

The central point of this is that there are infinite ways, flying being one of them, for us to explore what might be within us, to find what our potentials might truly be. An artist of any sort knows this. He knows that a sense of one's self, one's being, can be found through the serious and joyful practice of the art, be it music, painting, dance, or athletics, anything that demands skill and absorption. The art becomes the means of giving form and direction to the exploration. Of course, we can never find the object of our search unless we look for it, but before that we must understand that there is a question. The question, of course, is what might lie beneath the surface that we aren't aware of, that we don't yet comprehend. Even the fact that the question itself exists can be a revelation. How often do we stop to consider that there might be more than what is assumed by our ordinary consciousness? Do we ever pause to wonder what might be within ourselves or others, obscured by our own lack of vision and imagination? Somehow we need to have a sense that the moon is not full, that there is more.

Saint-Exupéry had this sense and a poet's grasp of what man could become. Once he was on a train that was crowded with laborers and their families returning home across Europe after working for years in France. Their lives were by necessity crude and harsh, with little time or thought given to more than the basics of physical survival. Saint-Exupéry was tormented by the deadened, truncated lives he saw, lives broken and anesthetized by an existence confined by the traditions of a restrictive society. His deepest torment was triggered by the sight of a child whose identity was yet to be formed, who was still unfolding, undefined. "This is the child Mozart. This is a life of beautiful promise. Little princes in legends are not different from this. Protected, sheltered, cultivated, what could not this child become?" He unhappily concluded that "This little Mozart is condemned. . . . What torments me is not the humps or hollows nor the ugliness. It is the sight, a little in all these men, of Mozart murdered."*

At least we are becoming aware that we have the chance to live our lives more fully and deeply than we may have thought possible. The things we have been speaking of, the different

---

* Saint-Exupéry, *Wind, Sand, and Stars*, pp. 305–306.

modes of consciousness, the self-actualization, the peak experiences, are all part of being human. They are not the stuff of the supernatural or the occult, not part of some cosmic mystery known only to clairvoyant oracles. They are simply part of all of us, part of our natural potential as human beings. Flying is just one of the ways of reaching for them. Since flying can be a path to lead us to much richer lives, it's disappointing that many others, even some who fly, can't think of it in this way. Perhaps they have yet to see the full moon and learn that there is more to flying, and to themselves, than they realize.

Much of the current writing about aviation directed toward an audience unfamiliar with flying seems to miss this central point. On what appears to be a regular schedule Sunday supplements include articles on flying. On a dare, an editor's assignment, or a long-gnawing ambition, someone learns to fly and discovers the "joy" of flying. This revelation, usually couched in "there I was at 10,000 feet" language, is the most common theme. The author depicts flying as great fun, but more importantly as therapy, a way to unwind and work out the knots of tension created by the burdens of everyday life. "Lots better than a psychiatrist's couch," to paraphrase a recent example. "What a fine way to get away from it all, if just for a while, on weekend flights to places well beyond the limits of freeway traffic. Really be able to face the world on Monday morning with a set of recharged batteries." And so on.

This is all true, of course, but it is a shallow kind of truth because it is incomplete and misses the essence of what flying is about. Therapy itself isn't a solution to anything. At best it is a remission, a temporary easing of pain and anguish that makes the load of physical or mental trauma bearable while the cause of the trauma remains. It is a means to an end, not an end in itself. If the therapy gives relief and allows the patient to gather strength to confront and eliminate the source of his trauma, it is successful. If it simply masks his symptoms for a while, it is not.

Flying can be an end in itself, as we have seen. Its joys are things of the moment, to be treasured because they exist for themselves, not because they lead to other rewards. It doesn't need to be rationalized in terms of what it can do for us after we have landed. The insights we have gained about ourselves and our culture are not bits of practical knowledge that are going to help us cope. In fact they may make coping even more difficult than it was before, as we will discuss later. That we have had these insights might just shed an entirely new light on our lives, our society, and the relationship between the two. At the very least we are going to be faced with some important questions about society's values and our own.

There is another faulty image of flying, reflected in the airplane manufacturers' advertisements in the general aviation magazines. The ads for Piper, Cessna, Beech, and others are scattered throughout the magazines along with ads for American and European sports cars. It is instructive to notice how they all seem to appeal to the same values. The ads for airplanes and sports cars alike pitch the familiar language and imagery of luxury, status, efficiency, and sexuality. It's as if the symbols of consumerism that are valid when associated with automobiles could be applied to airplanes, and were completely interchangeable. To some extent they may be, but the pleasures of flying lie at a much deeper level than the models in the ads could ever imagine. The ads overtly stress consumption as a means of achieving a kind of freedom from care and responsibility. They ignore the actuality that this kind of freedom is based on rapidly induced obsolescence, superficial values that judge a person by his material possessions, and the paradox that one probably has to tie oneself to society's treadmill in order to sustain it. A strange and limited kind of freedom indeed. And since flying has shown us a reality we hadn't known before and has caused us to cast some doubts on the validity of the ordinary reality we had always accepted, there is a delicious irony here. Advertising, which plays such an important role in defining reality for us in our consumer society, is in this case acting against its own best interests. It exhorts us to buy airplanes and to fly and thus collect the rewards of wealth and status, unaware that if we *do* fly we are likely to reject the values upon which advertising itself is built.

But here we are in our airplane, turning circles in the sky high above the rest of the world, in a new world we have discovered both within and without ourselves. Our airplane is safe and dependable, and will perform as we wish it to. It is well engineered and carefully put together and represents a high level of technological achievement. In keeping with our democratic traditions it is available to ordinary people, to you and me. No one can claim that flying is inexpensive, but it is certainly not limited to the wealthiest among us. It is another irony that the airplane, a machine that so represents the values of our society, should be the means by which we escape the imperatives of society and discover important truths about it and ourselves.

What are we going to do with this knowledge when we are

back on the ground? We may not be able to share it with others, not easily at any rate. What is the world going to seem like to us? We had better consider this before we return to earth.

N2175M

4. *The Descent*

Flying as Play

THIS is the last of the four basic maneuvers, and it can be flown just as smoothly and precisely as the others. Instrument pilots must be able to control their rate of descent skillfully to keep on an electronic glide slope when landing in fog or through a low ceiling. If the pilot who flies under visual flight rules plans carefully, he can arrive at the airport's pattern altitude from cruising altitude so gently that his passengers will hardly know they've descended. To descend, we simply reduce the power of the engine or point the nose slightly downward and adjust the controls so that the gravity acting on the airplane is greater than the lifting force provided by the wing. When the rate of descent and the airspeed are stabilized and held constant, all the forces of thrust, drag, weight, and lift acting on the airplane are again in balance.

We need to plan and prepare for our return to the earth as thoroughly as we planned our departure from it. Since we will be arriving at an airport where there are likely to be other aircraft, we must keep an even more careful lookout than we did while flying away from heavy traffic. Before we approach the airport, we should know the pattern altitude and anticipate the kind of entry we will probably make. The more information we have now, before we enter the vicinity of the airport—about winds, altimeter settings, runways in use, and other aircraft in the area—the better our planning can be. Once we have a plan in mind—even a tentative one that we abandon later—our flying becomes simpler and safer. We can concentrate on making a smooth and easy approach and landing, confident that we have all the information we need and have assembled it into comprehensible shape. The airplane itself must be made ready to land. The fuel selector is switched so that the engine draws gasoline from the tank with the most fuel remaining in it. If there is an electric auxiliary fuel pump, it is turned on; if for some reason we have to abort our landing when close to the ground, we will have to open the throttle and get full power from the engine. A sudden interruption in the fuel flow would be disastrous. If our airplane has retractable landing gear, it certainly should be extended before we land.

All these items and a few others should be on a checklist that is at hand or printed on the instrument panel so that the pilot needn't depend on his memory to recall all the steps. If he must rely on his memory, he can turn again to a handy acronym. The introduction of aircraft with retractable landing gear coincided with the popularity of a newspaper comic strip to produce "GUMP," a prelanding checklist that is still in use:

G—Gas switched to fullest tank
U—Undercarriage down
M—Mixture rich
P—Propeller at highest RPM

With a plan for landing in mind and the airplane ready, all we have left to do is to prepare ourselves for the transition back to earth. It's too bad we must start our descent and end our flight because, in addition to everything else, flying is a lot of fun. We may work very hard to learn how to fly and to sharpen our flying skills, but we also expend great amounts of energy in the process simply because it's good fun. Flying becomes a form of play for us.

In our ordinary lives we commonly distinguish between work and play. In our culture work is thought of as something we go to on Mondays to Fridays; play is what we do on weekends and vacations. Work and play are thus neatly separated in time and space, as well as in conception. We work so that we can afford to play; we play so that we will be better able to bear up under the strain of work. Each is seen not as an end in itself but as something that has value only in relation to something else.

But real play is its own reward, and generally the more physical and psychic energy devoted to the activity of play, the deeper and more gratifying the satisfaction it brings. One reason for this is that the world of play is somehow set apart from the regular world. Time and ordinary affairs are suspended while we are at play and the game is run out. Because we come to our own particular form of play voluntarily, simply because we want to, it is an expression of our freedom, an attempt to remake our world. While we are at play new rules apply that are different from those that regulate our lives at other times. Play also has a definite beginning and end that set it apart from ordinary time. I have several friends who spend all the weekends they can manage every winter fishing for steelhead salmon in the coastal rivers and streams of northern California. Their routine is not the same as the person who flies on Sunday in order to be refreshed on Monday. The fishermen spend their nonseason weekends in very ordinary ways. Saturday and Sunday are part of an accepted rhythm of life. It is only in the winter that they take to the streams, up before dawn to stand waist deep in icy, swiftly moving water, and they count themselves lucky if they get even as many as a dozen strikes in a season. Meals are forgotten and physical discomfort, if noticed at all, is ignored. Like a pilot on a flight, they are completely immersed in the moment, the here and now of what they are doing. Several seasons may pass between actual catches of the elusive steelhead. One friend doesn't even like to eat fish, so when he lands one, he generally releases it back into the stream, aptly illustrating that it is the intrinsic value of fishing itself—not the promise of bringing home a fish—that is important and that draws these people to the chilly waters.

Play is a natural part of life, and to deny it importance is to deny an aspect of our humanity. Anthropologists tell us that the drive to play is as basic to our nature as the drive to learn and that the two are closely related. Young primates, human or otherwise, spend a great deal of time at play of some sort. They play because it is a natural way to learn behavior patterns that will be useful when they are adults—and because play itself is enjoyable. The play may be directly imitative or richly embroidered with fantasy, but it is one of the important ways the young learn to be adult. Our culture, with its residual elements of puritanism and its down-to-business approach to life, tends to dismiss play as trivial and inconsequential. Play is for children—when they aren't at their lessons. Our schools have forgotten, or have never learned, that man is a playing animal and that play is an essential way to discover ourselves and the world around us. We have only to watch a child mature from infancy to realize this truth, but somehow as a society we seem to ignore it.

When we approach the matter in this way, it's obvious that flying can be a kind of play. We're up here in the air because we enjoy it and because it brings us a deep sense of pleasure and satisfaction and maybe because it gives us a rich panorama on which to project our fantasies. In *Climb for the Evening Star*, Tom Mayer tells of demonstrating lazy eights during his commercial pilot examination flight. The lazy eight is a difficult maneuver, a lovely combination of climbs, turns, and descents. With constantly changing pressures on the controls, the airplane's imaginary paintbrush draws a symmetrical figure eight, laid on its side, across 180° of the horizon. Mayer felt that he had done the maneuvers well, but when he finished, the flight examiner took over the controls and did some lazy eights himself, all beautifully. Flight examiners fly with a candidate for a license to evaluate his judgment and flying skills. Although they are ready to take the controls in an emergency or if the candidate becomes hopelessly confused or frustrated, they are actually along as observers, giving the candidate the opportunity to do his best without interference or assistance. But like a batting coach watching talented rookies in spring training struggle through the paces he mastered years before, a flight examiner sometimes can't resist the urge to take his turn at the plate. Forgetting this for the moment, Mayer sensed disaster when the examiner took the controls, thinking that he had failed the examination. "You did fine," the examiner said as he returned the controls. "But I like to fly too."* Richard Bach extended the image of flying as play and fantasy even further on the journey he describes in *Biplane*. In the early 1960s Bach purchased an antique open cockpit biplane on the East Coast and was faced with the problem of getting it to his home in southern California. He turned problems into play by deciding to make the trip as a 1920s barnstormer might. The beginning and end of his journey marked a time in which he was free to live by a set of rules of his own making, one that demanded he navigate by road maps and by following rivers and railroad tracks, landing in open fields instead of airports, and sleeping at night under the wings of his biplane. The book is a chronicle of his attempt to create his own time warp and to live for a while in a world that doesn't exist anymore. Is there any pilot

---

* Tom Mayer, *Climb for the Evening Star* (Boston: Houghton Mifflin, 1974), p. 10.

who hasn't some time become the Red Baron or a World War I ace, even if for just a moment? I doubt it.

Why the World Can Never Be the Same Again

EVEN as we live out our fantasies, the world we left a short time ago when we started our flight still exists, and we must return to it. While we have been in the air the world has been going on about its ordinary business and probably hasn't even noticed that we've been away. We need to be ready for the world to seem different for us, even though it is still the same as when we left it. It's we who have changed. Our flying has helped us learn things about ourselves that we didn't realize before and has given us the chance to experience ourselves and the universe in ways we had not known before. The end result is that we will go through a kind of culture shock similar to that which people feel when they return home after a long stay in some very different society. The chance to live outside of their usual surroundings and usual way of doing things can give them powerful insights into their own society. The ways of life and values that they had learned to accept as universal, as aspects of life and society that were to be taken as givens, suddenly seem capricious and irrational. They return to find that many of society's rules and practices are arbitrary, not reflections of universal truths. The traveler has seen for himself that human beings can live out the rhythms of existence to a melody different from the one he was taught was the only one possible. For a while he even adjusted his life to a new tune. The readjustment to his own society can be awkward and difficult because the returning native is aware of aspects of his society that its homebound members can't imagine exist.

Flying has made us aware of many elements of our world that those who stay on the ground don't perceive. Let's first consider some simple examples. The airport I fly from is built on filled marsh land on the west side of San Francisco Bay. A mile north of the runway there is a perfectly square lake fashioned by the contractors who built a subdivision, also on filled land, that features backyard access to sailboating and swimming through a series of canals that connect to the lake. The canals enter at the corners of the lake, giving it something of the shape of a diamond when seen from the air. It's an obvious landmark for pilots and for airport controllers, who use it as a guidepost for directing traffic in the pattern. "No turns until reaching the diamond-shaped waterway" is a vital rule for us to follow, and the diamond-shaped waterway is a vital part of our consciousness. There is a major freeway that runs along the bay shore that connects the communities along the peninsula with the city of San Francisco itself. The freeway is only a few hundred yards west of the runway and the diamond-shaped waterway. Thousands of people go by every day on their way back and forth to work, and probably very few of them know about the waterway because it can't be seen from the freeway.

There are signs that point out the exit ramp leading to the airport, and the tower can be seen from the freeway, but there are no clues that would lead a commuter to suspect that there was a lake so close by. Even the sailboats are hidden behind a row of trees. How remarkable that something so important to those who fly is outside the awareness of others who pass just as near.

The physical scale of the town I grew up in is nothing like I had thought it to be, but I never imagined that my perceptions might be faulty until I flew over the town on my way to somewhere else. I found that the house I lived in was really quite close to the school, not at all like the long walk I remember. I also noticed that there is a small reservoir behind the hills outside of town that I never knew was there. Now, when I go back to the town for a visit, I can see through the hills and picture the reservoir in my mind. But my family and friends, who haven't had the opportunity to examine their own turf from the vantage point I have, still don't know about the reservoir.

Flowing from a change of perspective, these insights can be gained by any observant airline passenger. They are instructive in helping us to understand more fully the physical dimensions of our world, but they are only shadows of more profound truths that flying can show us about the social world in which we live. We know, through the intensity of our own personal experience, that we are now able to perceive reality in ways that we hadn't learned about or perhaps hadn't even known about in our ordinary life on the ground. We also know, at the deepest levels of our being, that we are capable of accepting complete responsibility for our own physical survival. Within any given set of conditions that might influence a flight, we are able to judge for ourselves if our flying skills and our knowledge are strong enough for us to be able to cope successfully with the situation we anticipate. Within the broad range of possibilities that the laws governing aviation allow, we set our own limits, apply our own judgment, and act on the decisions we make. No one else makes our rules, no one else measures the myriad details that influence a flight, and no one else weighs their significance. In this process of taking charge of our own rationality, applying our highest capacity for thought because the stakes are so high, we are in turn taking full charge of our own lives.

Because it is our own judgment at work, our own decision to be in the air and rely only on ourselves, we are for the time we are flying living life at its fullest, pushing our humanity to its limits. Not only are we using all the powers we have, but we are gambling that they will be enough. Our bet is on ourselves and it is not covered. There is always the chance that we can be wrong. We do this all by ourselves, and we wager our existence on our trust in our own decisions.

What a revelation this is, to know that we are capable of such audacity. Within the common framework of our society, no one is assumed to be able to be so self-reliant. It's difficult, if not impossible, to learn this fact about ourselves in our ordinary life because ordinary life doesn't assume that it is possible. The paths our lives are expected to follow are designed to be shared, with all of us helping each other to hold to a common course. We must cooperate with our fellows for any human society to function, but we must realize that the path, the frame of reference that is created by society, is arbitrary, a convenient construct created and maintained to make living together possible. Since one of its purposes is to give stability to existence and to provide a commonly agreed on picture of what the universe is like, it must of necessity be limited. To lend order and predictability to life, some "givens" must be established and accepted, even at the cost of restricting the range of possibilities.

Flying has given us the chance to see beyond the frame of reference and learn what might lie outside of society's definitions of reality and of human possibility. Sociologist Peter Berger calls this phenomenon of seeing past society's network of assumptions "alteration," the acquisition of a new perspective on society. This sense of alteration—discovering that society is different from what we had always taken for granted—can come in an explosion of recognition or as the result of a long series of experiences. In any case we are confronted with the realization that we can't accept society at face value. Society becomes not something concrete and real, but a collection of façades that are supported by our conscious or unconscious willingness to consent to them. Once we recognize this fact our lives are changed forever. Berger gives a vivid example of what alteration means to him. Imagine walking down a street in a heavily bombed city at the end of World War II. It is a familiar street to us: we know each building, each front door, each window. We notice that the front wall of an apartment house has dropped away, revealing the rooms and the furniture of the people who live inside. Life may return to normal some day and a new wall may be put up, but for us the apartment house will never be the same. We have seen what is inside, what life is like behind the wall. Others will be only aware of a rebuilt apartment house since that is what is considered as truth, but not us. For us the new wall is invisible because we know what is behind it.

A burst of X-ray vision like this can be the source of the kind of joy that is part of the intensity of a peak experience. Suddenly the little light bulbs appear over our heads, and we have a great sense of "ah ha, so that's the way it really is." This is one reason why the story about the pilot lounge was such a powerful incident for me. The special door had nothing special behind it. I was caught up in trying out a new role that I had worked very hard to master, only to find out that the role wasn't important at all. My new skills and knowledge got me to the airport, but my new role got me to the same roomful of indifferent clutter that waited for anyone who chose to enter it. What a grand joke.

The child who learns how presents really get under the Christmas tree experiences alteration, although he may not find any humor in the situation. What he had believed in is revealed as a loving hoax, designed to help him share in the joy of the season, perhaps, but a hoax just the same. His younger brothers and sisters may still set out cookies and hot chocolate for Santa Claus, but he is in on the game and can see past what is reality for them, and St. Nick can never be taken seriously again. As adults this is a social fiction most of us are willing to go along with because we take it to be benign and harmless. Besides, it is directed at children, not at us.

There are social fictions directed at us that can be harmful—if we don't recognize them for what they are. Consider the way you and I are pictured by society in terms of accepted reality. In today's world individuals are usually defined by their occupation. In spite of the various movements and appeals to common decency to accept individuals as unique entities, each with his own special combination of qualities, our bureaucratic culture demands that we be defined as salesman, nurse, manager, or, the very worst of all, unemployed.

For a number of years I was an English teacher, an occupation that lays claim to a certain amount of shabby gentility in our culture. I can't count the times that on meeting a person for the first time and telling my occupation the new acquaintance would give a response that would always make me wince: "English teacher, huh? Guess I'll really have to be careful about my grammar when you're around." Is that what they saw in me? A cranky authority figure who quibbles about dangling participles and misplaced modifiers? One who takes seriously the things people in the "real" world dismiss as trivial? It was always difficult, and sometimes impossible, to overcome this stereotype. I understand its origins and could easily rationalize its existence, but I was always offended and felt a sense of personal resentment when the stereotype was applied so automatically to me. Why did people insist on forcing me into that pinched and narrow little category? I was much more than that. Whether one was attracted or repelled by all the other parts that went together to make up me, the mixture of good and evil, was irrelevant. The point is that there was more than could ever be contained under the label "English teacher."

This fiction is harmful to the individual, ourselves, and others. As in the example just given we are apt to lose the joy of experiencing the real person hidden behind the label that society has so neatly pinned on him. Also—and this has the potential to be personally devastating—we can accept the label we have been given and can assume the role that has been defined for us. When I worked for an engineering consulting firm, many of my colleagues were men recently retired from the military. In the military one's role is tightly defined in a rigid hierarchy. Everyone knows his place within that hierarchy and learns what is expected of him and what behavior is not tolerated. Social relations are regulated by the relative positions of individuals within the system. It's very easy after twenty or perhaps thirty years to not only assume but to become the role that was assigned.

That was the case with these men. They could only think of themselves in terms of the roles they had supposedly abandoned when they had retired. They related to each other in terms of their former ranks and technical specialties and spoke the jargon of the world they had known, as if they had never left it. Their identities had been established for them, and they knew of, or could conceive of, no other possibilities for themselves.

They had taken their new jobs because the company's principal client was the same branch of the military in which they had passed their careers, but the old roles were so deeply set that most of the men could not adjust to the new roles they were asked to play. They were not able to establish enough psychological distance between the military and themselves to be effective consultants. Although they were generally competent in their specific areas of expertise, they were too used to taking orders and not used to giving advice based on their own personal authority. Authority had always been bestowed on them by a superior; it was not something that came from within, based on intelligence and experience. Their attitudes about themselves were reinforced by the military people they were supposed to advise, especially those of senior rank, who saw the consultants as underlings still in uniform, subordinates whose function was to do their bidding and to accept and act on the uninformed bias of the most senior person. The consultants, of course, were not capable of doing anything else, and the project continually floundered as we waited for new instructions from those *we* had been hired to advise. I left the company before the fate of the project was determined. But the direction of the lives of the men I worked with seemed as fixed and as firmly controlled by forces outside of themselves as if they had decided, in middle age, to sign up for another tour of duty.

Slavish devotion to externally imposed roles is often the stuff of comedy. The operettas of Gilbert and Sullivan, written a century ago, still amuse us because they let us laugh at the absurdity of the captain who must trade places with a deckhand because of a mixup when they were both infants or of the young man who must turn against his fellow pirates because he has reached his majority and must now become a respectable citizen. But it is also part of tragedy. When the world no longer needed Willy Loman as a salesman, he could only respond with suicide. Willy could only perceive of himself in the deluded fantasy he had built around his existence, a role he filled to the extent of anticipating the kind of death he imagined a salesman should meet. When the illusion closed in on him and shattered, there was nothing left for him, nothing to hold his life together. His role was over and so was he.

When we really begin to analyze the situation, there isn't very much about society that isn't subject to question. Just about every aspect of our social world, its conventions, traditions, institutions, is apt to be different from what it first appears to be. No gigantic conspiracy of deception need be directed against us for this to happen. It's only that flyers see what others don't see. What others honestly accept as opaque we know to be transparent. Let's not let ourselves be trapped into a fit of ranting paranoia, but let's be considerably more skeptical than we have been and not take anything about society for granted, as "given," anymore.

There's nothing wrong with social fictions as long as we recognize them as such. They can help us to avoid unnecessarily hurting others, to protect ourselves, and generally to make life flow more smoothly and pleasantly. I engaged in a social fiction of my own when I first started developing an outline for this book. I was at work, bent over my desk, apparently very busy at my job. Protected by my look of industry and handwriting that is practically illegible, even to me, no one in the office could tell what I was doing, and in fact no one seemed to care. Certainly no one asked. My co-workers all seemed to be very busy themselves and engrossed in whatever they were doing. I had no idea what they might have been doing and didn't bother to ask. There's much to be said for the reciprocal courtesy of observing each other's fictions.

It's just when we knowingly accept and act on a fiction and try to pretend to ourselves that it is truth that our lives can go out of balance. We may even create our own fictions to protect ourselves from facing something that we truly know to be destructive or evil. In either case we screen ourselves from what we know to be true so that we can absolve ourselves of any responsibility for what might happen. Thus the merchant who overcharges for defective goods can proclaim caveat emptor, the politician can accept a bribe and call it a campaign contribution, and the citizens who lived near the Nazi death camps could feign ignorance of the real purpose of the chimneys.

We've had some powerful experiences in the air, experiences that have helped us see what we might really be. We have also seen that what we know to be true about ourselves might not match what society considers us to be. Further, we have discovered in more than one context that society's notion of reality is incomplete and that there is more to life and to the universe than conventional wisdom can contain or imagine. What now? Are we going to ignore what we have learned and go on about our lives as we have in the past as if nothing has happened? Are we going to pretend that we can deny our own knowledge? Is it possible that we can drop our flight down some Orwellian memory hole and make it disappear from our consciousness? I don't think so. The price we would have to pay for self-negation would eventually prove to be too high. We can't deny what we know is true.

When we took off, we committed ourselves to flight. Once we were in the air there was no turning back. We had no choice but to continue our flight until it ended with a safe landing. We might change our mind about our destination or about how long we want to remain in the air, but we can't change our mind about the flight itself. Our only choice is to deal with the flying. There is no limbo of indecision or ambivalence where we can go to sort things out and think about our situation. The single choice we have, unless we choose death over life, is to depend on our own skills, intelligence, and good judgment to see us through the flight and to bring us safely back to earth.

The Need for New Choices

Now, on the way back to the ground to end this flight, we must be ready to deal with the world from a new perspective. Just as we couldn't withdraw from flight once we were airborne, we can't back away from reexamining our lives in light of our recent experience. It doesn't have to be, but this has the potential to be more terrifying than any moment of fright we might have ever had in an airplane. We have to reconsider our entire existence in terms of our new knowledge. What can we keep? What must we change? What just won't work anymore and must be put aside? What new elements should we include? We are really plunging into the unknown here because no one has ever tried to sort out the universe from our own perspective before. No one ever will again, so our guidelines must be carefully established, although they can be changed later if we wish.

An excerpt from Barbara Tuchman's *The Guns of August* hinges on the need for guideposts. Unknowingly predicting the massive changes that World War I would bring, a British aristocrat wrote in the early 1900s, reflecting on the death of Edward VII and the sense of loss felt by the upper classes:

"There never was such a break-up. It was as if the buoys that marked the channels of our lives had been swept away." That is rather like our situation, except for one extremely important difference: we don't have to remain passive as events sweep over us; we can actively set out the buoys in the direction we choose as we explore the uncharted waters of our lives.

What we are speaking of here is truly taking responsibility for our own lives on the ground, just as we found we were able to do in the air. Since we can trust ourselves in the air, there's no reason why we can't apply that same self-confidence to our lives on the ground. Once we can see past society's "reality" and can begin to fashion a reality of our own based on our new knowledge, the possible directions that our lives might take seem infinite. There is no longer any inevitability about the course of our lives. The future is in our own hands, and we are not bound by a social time line that turns our lives into a linear series of predictable events from birth and childhood to retirement and death. Inner authority and responsibility have replaced external imperatives, and we are free to live within the dimensions that we set for ourselves, not those determined by others.

Any prediction of the future has to be grounded in the past and the present. Once we identify a trend or pattern of events, we can project it into the future and determine its possible strength. Thus we read descriptions of what America will be like in the year 2000, all based on projections of conditions that exist today. Only what we are aware of now can be considered in our projections. There is no way to include the unknown in our equations. This is why the pattern of our lives probably won't match what society might predict and assign to us. One of the assumptions implicit in any prediction is that a common notion about the nature of reality that exists today will exist tomorrow. Since we are doing some serious tinkering with reality, any prediction we make is thrown off-base and probably won't hold. We won't easily fit the production line model of life or follow the curves on actuarial charts. Our new data have blown the program. It won't compute.

This doesn't mean that we suddenly have the license to exploit and run roughshod over others. Although we may reject part of society's dictates, we are still social creatures who must share love and compassion with others if we wish our lives to be full and happy. It's only that we are trying to find our own definition of "full and happy." There's no one to tell us what will make us happy and give our lives meaning and then hand over to us whatever that might be. Even the Wizard of Oz turned out to be just a frightened old man, hidden behind a curtain, pulling levers and turning dials. The old fraud was wise enough, though, to show the pilgrims who discovered his secret that each already had the inner quality he wished the wizard could bestow. All we really know now is that we must somehow put into play all the powers and capacities that we have as human beings. What form this may take is up to us. We have the privilege and the responsibility to make the choice.

Since we have come to the conclusion that society's picture of reality is no longer valid for us, we need to create a structure that will make sense for us and will take into account the new information we have acquired. We can't abandon completely the old since it will be our point of departure. Besides, we must accept that we are in many important ways products of our cultural heritage and can't easily slough off the history, values, and traditions into which we were born. Nor would we want to. Our new reality flows from the old and is rooted in it. We just need to shift and adjust the old reality so that our new frame of reference will accommodate and be consistent with our new set of facts. This is what is called, in the history of science, a shift in paradigms—a paradigm being a general theory or set of assumptions into which observed data are fitted and related. When data are found that are not compatible with the traditional paradigm, a new framework must be shaped that will accommodate the new facts and still give a valid explanation of previously determined facts. Existing knowledge must be considered in light of the new framework, and some assumptions about that knowledge will probably have to be set aside as no longer tenable. Setting aside a familiar set of assumptions is a difficult and painful business since it may mean that we have to question not only the conceptual scheme at hand but other aspects of our lives that we thought were keystones of existence, and now seem to be assumptions themselves.

Sometimes it is less painful to twist and bend the new evidence to reconcile it to the old framework or to simply deny the existence of the evidence. Consider the reaction in Western civilization when Copernicus and others proposed a new paradigm concerning the relative motion of the earth, the stars, and what we now call the solar system. The old framework assumed the earth was the stationary center of the universe and that all heavenly bodies studied by astronomers for centuries revolved around it. Man was the master of the earth, and by extension God's finest work was the focal point of the entire universe. Thus cosmology and religion were considered inseparable and intertwined. The new evidence developed by careful observation simply was not compatible with the old conception, and the Copernican theory, which put the earth and the other known planets revolving around the sun, was advanced. The new paradigm accommodated the facts, gave a coherent picture of the universe, and was ultimately accepted after several centuries of turmoil not only in science and astronomy but in religion and philosophy as well. If the earth is not the center of the universe, what is? Weren't all the celestial bodies created by God for the benefit of man? To accept the idea that the earth revolved around the sun and not the other way around meant that many other accepted notions about God, man, and the cosmos had to be questioned and reexamined. One's entire frame of reference, the bedrock of certainty that gives continuity and stability to life, would be severely tested. Resistance was formidable. Both Protestant and Catholic Christianity were particularly threatened, and religious opposition to the new interpretation of the universe reached its peak in the Church's trial of Galileo in the early seventeenth century. Even Galileo, who was seen as a giant in his own time and is seen today as one of the pioneers of modern science, could not stand up to such power, and was forced to recant the validity of his own observations and writings based on the assumptions of the Copernican heliocentric solar system.

Even today we have linguistic vestiges of the old conception. We still speak of the sun rising or setting even though we know that the sun stands still while the earth turns, that we passengers are just constantly changing our point of reference. Astronaut Michael Collins reports that it took a trip to the moon to break himself of this habit of thought.

Once we have established the broad outlines of our new

reality, our new paradigm, we can begin to explore its implications and possibilities. This can be an incredibly exciting adventure because there is no way to predict what we will find. Copernicus could not have anticipated that his conception of the earth as a satellite of the sun, combined with Newton's theories of gravity and the forces acting between bodies, would underlie the thinking that made possible Collins' trip to the moon. While we are engaged in the rather risky but fascinating business of working out our lives in the new framework, we need to remind ourselves that others don't share it with us. There will be many elements in common, but it's a reality we've created for ourselves. Some of the things we've learned we may be able to share, but unless others have had experiences such as ours that opened us to the possibilities we now see, we may not have a common medium through which to communicate. Those qualities we find in ourselves and in the world, qualities that are so clear and real to us now, may still be invisible to others. They may not see us as we are but see us still in the roles that we find no longer adequate. So we have to be patient with others and ready for a certain amount of loneliness. After all flying itself can become very lonely, especially at first, when we aren't yet aware of what its promise might be for us.

One Saturday morning when I was a student pilot in California I set out on a cross-country flight from Hayward to Ukiah, a distance of about one hundred miles. I didn't finish the trip because the weather in my path looked much worse than had been predicted—but I did get as far as Berkeley before I turned back. Since it was early Saturday morning, I knew that many of the people below me were still asleep or in bed trying to shut out the world for a few more minutes of rest before starting the day. If I could have taken the time to look, I could have seen the dormitory where I lived when I was a college student. I knew how quiet it would be there and how few people would be up and about. For the first time while flying solo I had an overwhelming rush of loneliness as deep as I had ever felt before in my life. My solitary world in the tiny cockpit seemed to close in on me. Instead of opening the universe to me, flying cut me off from everything but what was inside that tiny airplane suspended 3,000 feet above the city. It was not a sense of fear—I was beyond that in the routine kind of flying I was then doing—it was a profound sense of loneliness. The weather was deteriorating and turning back was a wise decision. I can admit to myself now that the weather was a convenient excuse and that I probably would have turned back anyway. I simply was desperate for human company.

This points up another lesson that flying can teach us. Since we gain a deeper sense of who and what we are, we need others less to validate our existence. Since we are simply ourselves, not the stereotype of a role that we choose or is assigned to us, we don't need signals of reassurance from others. What we do need is to be able to relate to others in ways that extend beyond social roles to others' uniqueness as human beings. This kind of natural compassion can come when we are secure in ourselves and confident that we don't need superficial relationships to stave off loneliness or insecurity. Relationships between people rather than roles are worth the effort of the search. In the meantime our own resources are more than adequate to sustain us.

One of the commitments we may make, now that we are free

to choose those that have meaning to us, might be to try to convince others that they have within themselves what we have discovered within ourselves. Conversations about these things are awkward and imprecise because of the lack of common experience and the difficulties of making language work effectively for us in these areas. Surprisingly enough it's sometimes difficult to discuss anything but techniques and gadgetry with some pilots, particularly younger ones, the men and women who dream of someday flying from the left seat of a 747. Their flying skills are superb, and their knowledge of the technology of aviation is thorough, but with only a few exceptions, that's where their sense of flying seems to stop. I suspect that at this point in their lives the image of an airline captain contains such power and magic for them that it overshadows the acceptance of the power and magic that is theirs for the taking.

Perhaps we should go on about our lives telling what we know to others who might be interested without sermonizing about some mysteriously revealed truth that only we possess. Lest we start taking ourselves too seriously, let's admit that all we have done is to become aware of a few things that our culture tends to cover up. At the same time that we learned about ourselves, we found out about our culture as well. We certainly can't pretend we are some sort of supermen because we have qualities all humans share, but all don't realize they have. We can't be diffident Clark Kents, either, pretending these qualities don't exist. Either would be a deception.

All cultures don't have the problems with individual identity that we do. In *Roots*, Alex Haley tells of a tradition in the village to which he traced his ancestry. After the birth of a child, the father would spend the next seven days deciding on a proper name for it. On the eighth day the villagers would gather to hear what name was selected. The father would make the announcement, but only after he had whispered the name into the infant's ear, the first time the name had ever been spoken as the child's name. The ritual was important because the people of the village believed that each person should be first to know who he himself was, only then was society told.

We are close to the airport now, and in just a few minutes we will have to end our conversation and devote our full attention to entering the traffic pattern and landing. Beginning flight students usually have a great deal of difficulty in finding their home airport when returning from a lesson with their instructor. They might be familiar with the area from the ground, but they haven't learned to recognize and interpret the clues needed to locate the airport from the vantage point of altitude, where the familiar takes on a new perspective. Once they have made a few flights, they generally develop a much better idea of where they are, both on the ground and in the air, because they have a keener sense of the pieces of their world and how they fit together.

Being able to find the airport from the air is an important first step in learning that the world holds more than we might have thought, and the landing itself is the transition between our world as it can be and the world as we have known it. A safe and graceful landing is an exacting maneuver: airspeed and rate of descent must be held precisely with smoothly coordinated adjustments of pitch attitude and engine power. Control must be even and sure, and the pilot must be absolutely certain of the effects of each move he makes since every landing is made under a different set of conditions. The factors that affect the airplane's performance—wind direction and velocity, air temperature and density, the airplane's weight—are never exactly the same, so the pilot's response to them can never be exactly the same either. To pass without incident through the boundary of his two worlds, he must perceive all these factors and establish harmony among them. When the pilot does this, his airplane gives up its ability to fly past exactly the section of the runway he selects: it touches down softly, and his flight is over.

And so is ours.

Epilogue

Now that we are back on the ground, we can try to put our flight into perspective. There's no point in taking another inventory of what we have learned. What we must do now is to decide what we can do with our knowledge. One choice would be to ignore it. We could just go on about our normal routine of existence as if nothing had happened and shrug off our experience as if it were some sort of dream—a provocative one, certainly, but a dream nonetheless. This is a possibility, of course, but to treat as a dream the idea of taking responsibility for our own lives on the ground—as we know we can in the air—might be too costly in the end. Better we never should have taken our flight and remained in ignorance than to suppress such potent knowledge. To do so would be an exercise in self-denial; it would mean walling up parts of ourselves and putting them out of reach, where they would never be used or given the opportunity of expression. Existing this way, consciously and deliberately alienated from part of ourself, is living a stunted life, one that does not integrate and bring to bear all one's potential to the enterprise of being fully human. Not only would we be cut off from ourselves; we would be cut off from others. By choosing to limit ourselves we would in turn be limiting the ways we have to reach out and touch the lives of our fellows, and our lives and theirs would be diminished further. If man is to be free to control his own life, he must be free to make his own choices—but to choose to live at less than one's full potential as a human being is to choose to live an impoverished existence.

Whether by conscious choice, lack of self-understanding, or force of circumstances, however, many people seem to live this way, even some who are regarded as successful by society and who are accorded power, prestige, and material wealth. According to Michael Maccoby, who studied the lives and attitudes of over 250 managers in a dozen major companies throughout the country, this kind of internal division and rejection of part of one's nature is common among the leaders of America's businesses and corporations. Maccoby found that these men tended to suffer from what he called an "undeveloped heart." Brought on in part by the intense competition among companies, the struggle for position within a company, and a dedication to little else than corporate and career goals, the managers had diminished capacities for compassion and feeling. Decent and moral men, they had detached themselves from their emotions in order to succeed, and success required setting aside thinking with the heart in favor of thinking with the head. Maccoby's head/heart distinction is analogous to the differences in left and right hemisphere thinking we discussed previously. Thinking with the head is using our rational powers to separate, define, and deal with information in a logical sequence; thinking with the heart is using the holistic, intuitive, and integrative mode of thought and perception. Since the imperatives of the corporation value abstraction and technique over emotion and feeling, these are the qualities that are rewarded. As a result of this severe detachment Maccoby found that most of the men had few deep emotional attachments to others and in general had difficulty establishing friendships that were little more than superficial. This detachment leads to isolation, and isolation to further locking oneself into the corporate role. Most of the men interviewed had a vague sense of malaise and had a semblance of an idea of what they might be missing, but few had solutions for correcting the imbalance, "the pathology of the heart." In any case these managers lost some of the richness that life has to offer, a loss to them as individuals, as well as, by extension, a loss to each of us. We should question the validity of a culture that insists that an individual surrender part of his humanity in order to achieve that culture's definition of success.

One of the functions of a human culture is to provide a reasonably stable and predictable environment that allows humans to develop and survive. It is a protective mechanism, giving its members a complex set of institutions designed to ensure the survival of the individual and the group. The diverse strands of the culture are held together by a commonly held definition of what constitutes reality. Although stability and consistency are important if the cultural context is to allow human growth to take place, there may be a point where stability atrophies into rigidity and, instead of enhancing growth, in fact tends to inhibit it. This might be the case with the managers. The culture grows even stronger than the individual it is supposed to protect and nourish, and man's creation turns back on him and works against the fulfillment of his potential. Surely we could never be aware of this if we had not the benefit of an experience like our flight, which gave us the rare chance to gain some knowledge of ourselves that we probably could not have discovered otherwise.

Before we totally condemn culture's attempt to remove all uncertainty in our lives, let's look at an alternative—albeit an extreme one. There have been times in human history when great chunks of a culture's accepted reality have seemed to fall away and disappear. When this happens, the people who had guided their lives by the promises of deeply rooted truths become frightened and confused, and feel deceived. They don't know, as you and I do, that a culture's reality is only a workable illusion, a tentative explanation of the universe that can't be accepted as absolute. In their frustration and resentment it is easy for them to be seduced by simple explanations for what must have gone wrong. Scapegoats are discovered who are blamed for destroying the now-shattered illusion that is still thought to be reality. The idea that a false reality was accepted as truth, or that pipe dreams of the future were believed, is never considered. How simple it is in times of great uncertainty to surrender ourselves to a person or a cause that promises security and a safe return to the familiar and the comfortable. No such pledge can ever be fulfilled—we know as much from hindsight—but there is always the chance that you and I could fall into the same trap. Nazism in Germany can be reasonably understood in these terms, as can the KKK and McCarthyism in America.

So the question we each must face in very personal terms is just how much uncertainty can we tolerate? We are now considering taking complete charge and responsibility for our lives, creating an existence that strikes a balance between what we know about ourselves and what we know about the universe—one that can absorb the jolts of new knowledge that are bound to come and disturb the stability we have established. Like every creative endeavor the outcome is not known in advance. We can't tell at the beginning if we are going to make this all-consuming project work or if it will come crashing down around our ears. We are moving into the unknown, into territories not already surveyed and mapped by our culture, and we need to be ready to deal with the terror of uncertainty. We just don't know. Let's not try to disguise our predicament with glibness. Even if others have traveled similar paths, we have not, and our fears are genuine. But since this is a conscious decision we have made, based on a greater measure of self-knowledge than we would have had if we had fallen into this situation by accident, we are forearmed. We have the luxury of time and intelligence that allows us to do some careful planning.

When engineers design complicated systems of any kind, they try to anticipate the most obvious operational problems the systems will face. Once the problems are identified, means of avoiding them or of reducing their impact are built into the original plan. One method used to foresee problems is the "worst case" technique. The most disastrous circumstances imaginable are postulated, and ways are devised to respond to the crisis and to ensure the safety and survival of the system and those who operate it. Our lives don't spring from an engineer's drawing board, but we can apply some of his technique to our future. What problems can we foresee? What is our own personal "worst case"? How clearly and calmly can we picture the details if the "worst case" actually happened? When we do this, when we locate and define potential difficulties in advance, the unknown slowly becomes the known. Once we anticipate, even to a small extent, what lies ahead, our anxieties will begin to fade. It's the fear of the unknown that really frightens us.

A "worst case" for a pilot is an engine failure just after takeoff, and that's when it happened to me. It was just the moment I had tried to prepare my students to anticipate. When the engine stopped, I went through the responses—unconsciously, automatically—that I wanted them to make. The potential disaster became a drill, a chance to carry out to its ultimate conclusion an often practiced exercise. The airplane was destroyed. My student and I climbed out, examined the wreckage, and watched with bemusement as the ambulances and fire engines raced to save us.

With our apprehensions about the future reduced or at least made manageable, we can put our imagination to a greater test. It's easy enough to create a "worst case" scenario, but what about a "best case"? Since we are going to redirect our lives, we ought to have a goal in mind, even if it turns out to be a shifting one. The problem is to be able to envision the best of all possible worlds. Again our culture tends to restrict us because we have been conditioned to accept the popular notion of the good life. Our task, and what a joyful task it is, is to make for ourselves the richest, most humane, and most deeply satisfying world that we can conceive. All that we have learned about ourselves and our present world can only guide us and give us a notion of the wide possibilities within our reach. Besides creating the general outline of our ideal world, being responsible for one's own life also means determining the specific ingredients that best fit into our personal scheme of existence.

Building a new world for ourselves is exciting to anticipate, just as flying is exciting in its way. The adventure in flying lies in the flying itself and in the prospects it can open for us. It has the capacity to shake us loose from the cozy certainties of ordinary life and plunge us into reconstructing the universe. While there are other ways to forge this reality, I know that flying is one of the most powerful and reliable.

When I was eleven years old, I pestered my father into taking the family for an airplane ride. We drove to the nearest airport, found a pilot who was willing to take us up for a fair price, and packed ourselves into his airplane. I'm not sure, but I think it was a Stinson. My father enjoyed himself, my mother almost got airsick, my younger brother was mildly amused, and I was ecstatic.

I was the only one among my pals in the neighborhood ever to have been in an airplane, so I got to tell the story of our ten-minute flight many times. The more times I told the story, the more I fantasized about what it might mean. Does being up in the air make us more susceptible to strange diseases and ideas? Does it make us immune to them? All my daydreams seemed to turn on one question: Does leaving the ground, even for only a short while, have the potential to make us somehow different from what we were before?

As I grew older I dismissed these thoughts as childhood fancies. Now I know that the child was right. It's true that flying has the power to change us if we so decide. Those who look deeply into what flying can be have the chance to make its ordinary magic their own. There's no secret or mystery; it's simply one of the ways we have to live a richer and more complete life. There is only a choice that must be made, and even if your flight above the ground has been an imaginary one, you have a choice to make as well.

Bibliography

This bibliography includes the works that most influenced the writing of this book. The sources of all direct quotations are indicated if not given in the text. Many of the books have extensive bibliographies of their own that can be used by the reader to further explore topics of interest. Where the exploration might lead is of course unknown, which makes it all the more exciting and important. The annotations attempt to present some idea of what each book is about, information about the author when appropriate, and, in some cases, questions that the book raises.

Adams, James L. *Conceptual Blockbusting.* San Francisco: W. H. Freeman and Company, 1974.
Ways to generate better ideas and avoid mental dead ends by a professor of engineering design. An excellent discussion of important literature on thinking and creativity is included.

Bach, Richard. *Biplane.* New York: Harper & Row, 1966; New York: Avon, 1973.
Bach's account of his barnstorming fantasy while crossing the country in an ancient open-cockpit biplane. See also *Stranger to the Ground* (New York: Harper & Row, 1963, 1973; New York: Avon, 1974), *Nothing by Chance* (New York: William Morrow & Co., 1969; New York: Avon, 1973), and *A Gift of Wings* (New York:

Delacorte Press/Eleanor Friede, 1974; New York: Dell, 1976), for perceptive reflections on flying, all based on Bach's own experience. More imaginative treatments of what flying can reveal about ourselves and the universe are found in *Jonathan Livingston Seagull* (New York: Macmillan Company, 1970; New York: Avon, 1973) and *Illusions* (New York: Delacorte Press/Eleanor Friede, 1977; New York: Dell, 1979).

Berger, Peter L. *The Precarious Vision.* New York: Doubleday & Company, Inc., 1961.
A sociological analysis of social fictions in modern life and the role of organized religion in maintaining them. Some fictions we engage in may be self-serving, while others may involve us unawares. Why did everyone but the child "see" the emperor's new clothes?

——— and Luckmann, Thomas. *The Social Construction of Reality.* New York: Doubleday & Company, Inc., 1966; New York: Anchor Books, 1967.
An important theoretical work in the sociological subdiscipline of the sociology of knowledge. Since reality is a human creation, it can be manipulated and controlled by particular groups of society for their own advantage. A difficult but worthwhile book.

Capra, Fritjof. *The Tao of Physics.* Boulder, Colorado: Shambhala Publications, 1976; New York: Bantam Books, 1977.
The author is a physicist who has studied the subtleties of Eastern philosophy, and his book lucidly explains the conceptual basis of each and the relationships between them. The implications of his work are enormous. If Taoism and physics, for example, are really so similar, what other ap-

parently contradictory aspects of our world might also be as closely related?

Castaneda, Carlos. *The Separate Reality.* New York: Simon & Schuster, 1971; New York: Touchstone Books, 1972.
The second in Castaneda's series of don Juan books. Also see *The Teachings of don Juan: A Yaqui Way of Knowledge* (Berkeley: University of California Press, 1968; New York: Pocket Books, 1976) and *Journey to Ixtlan* (New York: Simon & Schuster, 1972; New York: Pocket Books, 1976). Each deals with Castaneda's efforts to perceive and to accept the reality of don Juan's universe. One must read the books with an attitude of openness and curiosity.

Duke, Neville, and Lanchberg, Edward, eds. *The Saga of Flight.* New York: The John Day Company, 1961.
The Lindbergh quotation appears on page 199 of this anthology of writing about aviation and was taken from his book, *Of Flight and Life* (New York: Charles Scribner's Sons, 1948).

Gallwey, W. Timothy. *The Inner Game of Tennis.* New York: Random House, 1974.
A tennis pro's method of applying the techniques of Zen and right- and left-hemisphere thinking to athletics. His approach to tennis can be directly adapted to flying. See also Gallwey's *Inner Skiing* (New York: Random House, 1977).

Gilbert, James, ed. *Skywriting: An Aviation Anthology.* New York: St. Martin's Press, 1978.
A recent anthology of aviation literature. The choice of authors—Bach, Gann, Lindbergh, Saint-Exupéry, Shute,

and others—includes the finest who have written about the experience of flying.

Hall, Edward, T. *Beyond Culture.* New York: Anchor Press/ Doubleday & Company, Inc., 1976; New York: Anchor Books, 1977.
Hall, an anthropologist, marshals a wide variety of powerful evidence to show that Western man has created a culture that in many ways contradicts and denies his physical and psychological nature. The book includes a valuable bibliography. Can we invent a culture that would complement and enhance our biology?

Hayakawa, S. I. *Language in Thought and Action.* New York: Harcourt, Brace & World, 1949; 4th edition, New York: Harcourt Brace Jovanovich, 1978.
A classic work on semantics by the current senator from California. Some of the everyday examples chosen to illustrate ideas may now seem dated, but the ideas themselves are not.

Holt, John. *Never Too Late: My Musical Life Story.* New York: Delacorte Press/Seymour Lawrence, 1978.
Widely known for his writings on education—*How Children Fail* (New York: Pitman Publishing Corp., 1964; New York: Dell, 1967) and *How Children Learn* (New York: Pitman Publishing Corp., 1967; New York: Dell, 1972), etc.—Holt began to play the cello and study music seriously when he was fifty. The book is about the demanding discipline and the joy that music has brought to him. Why would a person choose such an endeavor? Why do some people choose to learn to fly?

Huizinga, J. *Homo Ludens: A Study of the Play Element in Culture.* Boston: Beacon Press, 1955.

Written more than a generation ago by a European historian, the book is still a major work in its field. Huizinga saw play as a vital part of human existence. Why can there be no virtue without cakes and ale? Does all work and no play *really* make Jack a dull boy?

Kuhn, Thomas S. *The Structure of Scientific Revolutions*, 2nd ed. Chicago: University of Chicago Press, 1970.

According to the author, the history of science is not the smooth progression from Archimedes to Einstein and beyond as reflected in the textbooks. Rather, science advances when the paradigm—the assumptions about the nature of the universe—of the scientist no longer can explain his objective observations of that same universe. A problem is giving up the old set of assumptions that had previously defined the range of possibilities. How difficult must it have been to abandon the search for a perpetual motion machine?

Lindbergh, Charles A. *The Spirit of St. Louis.* New York: Charles Scribner's Sons, 1953.

The story of the Paris flight, first published more than a quarter century after the event. The interval gave Lindbergh the perspective to better see the flight within the context of his life prior to 1927. What does one think about when so completely alone?

———. *Autobiography of Values.* New York: Harcourt Brace Jovanovich, 1978.

Edited and published after his death, the book is a collection of previously unpublished writings that attempt to chronicle an extraordinary life and to define man and his place in the universe. Rich insights into the complex mind of a gifted man.

Maccoby, Michael. *The Gamesmen.* New York: Simon & Schuster, 1976.
A description and analysis of men who hold influential management positions in American corporations. Some become so enmeshed in their professional roles that they lose sight of other aspects of existence. Somewhat frightening in its implications.

Maslow, Abraham H. *Toward a Psychology of Being*, 2nd ed. New York: Van Nostrand, 1968; New York: Van Nostrand Reinhold Co., 1968.
Maslow's basic statement of his theories on self-actualizing, human creativity, and peak experiences. A pioneer in what became known as the "humanistic" school of psychology, Maslow believed that the limits of human possibility have yet to be determined.

———. *The Farther Reaches of Human Nature.* New York: Viking Press, 1971; New York: Penguin Books, 1976.
A collection of writings published after his death, the book is a helpful introduction to the range and depth of Maslow's thought. It includes a bibliography of all his published work.

Mayer, Tom. *Climb for the Evening Star.* Boston: Houghton Mifflin Company, 1974.
A journalist and pilot writes of his flying experiences. Classic airplane buffs will appreciate the material about Mayer's Staggerwing Beech.

Novello, Joseph R., and Youssef, Zakhour I. "Psycho-Social Studies in General Aviation." *Aerospace Medicine*, February 1974, pp. 185–188, and June 1974, pp. 630–633.
Reports of scientific studies designed to identify personality differences between pilots and the general population. This and other works have shown that as a group pilots tend to have a personality profile different from the rest of the population.

Ornstein, Robert E. *The Psychology of Consciousness*, Second Edition. New York: Harcourt, Brace, Jovanovich, 1977; New York: Penguin Books, 1975.
A remarkable book that explains right- and left-hemisphere thinking and perception in cultural as well as physiological terms. Ornstein points out the linear/holistic duality in a variety of cultural and philosophic traditions and suggests that modern psychology should not limit itself to a single mode of perception. He implies that psychology is in the midst of a dramatic shift of paradigm as described by Kuhn.

Pearce, Joseph Chilton. *The Crack in the Cosmic Egg: Challenging Constructs of Mind and Reality.* New York: Julian Press, 1971; New York: Pocket Books, 1973.

———. *Exploring the Crack in the Cosmic Egg.* New York: Julian Press, 1974; New York: Pocket Books, 1975.
As the titles suggest, one book is an extension of the other. The cosmic egg is a symbol for the cultural reality into which we are born. Once we recognize that reality is a social construct, the cracks begin to appear. Then we are free to explore them if we choose. An optimistic view of man, developed through a synthesis of physiology, psychology, anthropology, and philosophy.

Sagan, Carl. *The Dragons of Eden.* New York: Random House, 1977; New York: Ballantine Books, 1978.
The evolution of human intelligence from the earliest forms of life to the present. Sagan defines man in terms of his biological development and includes a full explanation of right- and left-hemisphere thinking.

Saint-Exupéry, Antoine. *Wind, Sand, and Stars.* New York: Harcourt, Brace & World, 1940; New York: Harbrace Paperbound Library, 1962.
One of the finest statements on the true essence of flying. Saint-Exupéry, a French pilot who helped pioneer airmail routes in Africa and South America, had a hopeful vision of the strength and sensitivity of mankind. He wrote this account of his own experiences and several works of fiction, all still available, before he was lost on a flight over the Mediterranean in 1944.

Walsh, John Evangelist. *One Day at Kitty Hawk.* New York: Thomas Y. Crowell, 1975.
A carefully researched account of the Wright brothers and the development and successes of their flying machines. Set against the background of aviation's early years, it also sheds light on the personalities of these two gifted, if mildly eccentric, men.

Wilson, Colin. *New Pathways in Psychology.* New York: Taplinger Publishing Company, 1972; New York: New American Library, 1974.
The life and thought of Abraham Maslow by a highly sympathetic biographer.